D1097949
BLACK SEA
TURKEY
SYRIA
IRAQ
IRAN
(PERSIA)
TRANS-JORDAN
PERSIAN GULF
GULF OF OMAN
SAUDI ARABIA
HEJAZ
OMAN
RUB AL KHALI
RED SEA
YEMEN
ADEN
HADHRAMAUT
GULF OF ADEN
INDIAN OCEAN
GYPT
DAN
Palacios

THE FORGOTTEN ALLY

Also by
Pierre van Paassen

DAYS OF OUR YEARS

THE TIME IS NOW

THAT DAY ALONE

THE FORGOTTEN ALLY

By

PIERRE VAN PAASSEN

DIAL PRESS • NEW YORK • 1943

First Printing September 1943
Second Printing September 1943
Third Printing September 1943
Fourth Printing December 1943
Fifth Printing January 1944
Sixth Printing February 1944
Seventh Printing April 1944
Eighth Printing July 1944
Ninth Printing September 1944

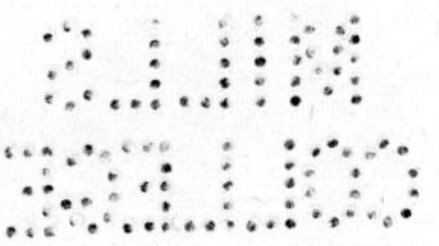

Designed by William R. Meinhardt. Manufactured in the United States of America by Haddon Craftsmen, Inc., Scranton, Pa.

AUTHOR'S PREFACE

Since the war now raging is one of global dimensions, involving the future and destiny of all mankind, it may appear to some that the solution of the problems of so small a country as Palestine, will be but an incidental, if not a task of minor importance for the United Nations at some future peace conference. I have nevertheless made Palestine the central theme of this book.

I have done this because I do not believe that the defeat of the Axis will automatically solve the Jewish problem with which Palestine is intimately and inextricably bound up. For that question is in fact being settled now; on the one hand by the frightful and uninterrupted killing of Jews in Europe and on the other by the lack of interest in high places, both in Britain and in America, in providing the terrorized remnants of European Jewry with a means of escape or even with a ray of hope for survival.

In the welter of immense and vexing problems facing the United Nations in Europe, Asia, and Africa, especially in the world's traditional areas of colonialism, wherein Palestine is located, the present method of dealing with the Jews by reassuring them from time to time with saccharine but perfectly meaningless phrases and on the other hand by enforcing their exclusion from Palestine may well be, I am afraid, indicative of the manner and spirit in which other peoples' problems are to be dealt with by the secret diplomacy of an unregenerate imperialism.

The fate of the Jews is being settled today by political events in the Near East and by decrees issued by Britain's Colonial Office. The new and better world that is to emerge from mankind's present agony is becoming visible in the Holy Land. It is not a happy or inspiring sight.

As one who is aware and who feels with a sense of personal involvement Christianity's guilt in the Jewish people's woes and the constant deepening of their anguish, I could no longer be silent. I made no emotional appeal. My language is not violent; the facts are. I simply stated the case of the Jews and their country for the consideration of all men of good will.

PIERRE VAN PAASSEN

CONTENTS

CHAPTER 1

THERE ARE NO MORE PROPHETS!

In my childhood in Holland—I may have been fifteen or sixteen years of age—I remember one evening accompanying my parents to church when prayers were offered there for the return of the People of Israel to the Holy Land. There was no haughty condescension or self-righteousness about those prayers. The congregation consisted of humble Calvinist shopkeepers and farmers, a cross-section of the *kleine luyden,* the small people of the Netherlands whose fathers had fought the Spanish Empire in a war of eighty years' duration for the sake of freedom of conscience. The Master of the Universe was implored to restore His People of Israel to the land He had promised them, because in the Christian world, throughout which they were dispersed they had, almost everywhere, been treated abominably. It was said from the pulpit that the sins of the Czar of Russia (for it was a time of pogroms in the Ukraine) were our sins. It was by our fault, the fault of universal Christendom and therefore of every individual Christian, that despite two thousand years of preaching the Gospel's lesson of brotherly love, the Jews could still not live in peace in our Western world. It was crucifying Christ anew, it was shattering His mystical body, said the minister, to allow the kinsmen of the Prophets and of the Lord himself to be hunted as wild beasts from one land to another . . . .

The return of the Jews to the land of their origin was not only asked in order that we might thereby be liberated from our "blood guilt" and our share of coresponsibility in the matter of their age-long persecution, but that the name of God might be glorified by prophecy fulfilled and by the ancient lamps on Zion's hill being rekindled as a light to all mankind. Every people had its own function and destiny. In order to fulfill that destiny and task the Jews must preserve their own national character, live again in their own house, and make of their Hebraic civilization a consciously rendered contribution to the sum total of the building of the Kingdom of God on earth, with its thousand gates and its scores of widely divergent national architects . . . .

After the service, the minister, a Walloon Huguenot by the name of Caspar Daubanton, stopped at my father's house and was lodged in my room. As we lay in the dark, I said to him: "But the Jews of Holland will never go to Palestine. And as to the others, in the other countries, how many will be left by the recurrent waves of terror to take up in Palestine where their fathers left off nearly two thousand years ago?"

Daubanton replied: "Do not say: they have left off. Judaism is still a living reality. It has been transplanted, it is true, from the Holy Land and has unquestionably suffered in the process of being torn from its roots, its locale, its soil, and its spiritual climate. But it is not dead. Since the Jews left Palestine they have created things of great spiritual value—the Talmud, for instance. The spiritual creativity of Judaism has not withered. Judaism is not dead. It suffers from anemia. It has no anchor. It would regain its health if allowed to function unhampered, and that can only be if the Jewish people have back a place where they can work freely."

And then referring to my remark about the Dutch Jews, he expressed the hope that they would never wish to depart from us. They had been an honor to Holland and true brothers from the day they came to us from Spain. And then Dr. Daubanton said something about which I have often thought in later years as his words came back to me. "There is," he said, "a mystery about the people of the Jews, a mystery which both attracts and repels us. Sometimes I think," he went on, "that the mystery resides in the fact that they, unconsciously perhaps, as a people, are the bearers of God's word . . . . We know they are . . . . We feel they are. And deep down in our hearts we hate them for it. For we hate God and do not want to follow His law. In their mere presence there lies always, I find, a subtle, often unavowed and undefinable challenge to us, something of a reproof, an accusation. They remind us of something of which we do not like to be reminded. I think," he said, "that there sits the core of the secret of Jewish persecution in Christian lands."

And then he went on to speak about the dread possibility of that mentally reserved hatred awakening some day in Europe. "Some incident, insignificant in itself," I remember him saying, "may very well bring it into the open. Not long ago we saw it come snarling to the surface with the Dreyfus affair in a country as civilized as France. Who knows when and where it will break out again? And so long as that hatred persists in our hearts," he said gravely, "the Jews are not safe anywhere, not in Russia, not in Germany, not in Holland, or in that far-off America to which so many of them are fleeing these days from Russia. In moments of historical crisis they are first thought of, almost instinctively, as the convenient victim on whom to unload popular frustration and wrath or fury at governmental incompetence, as we see it happening today in the empire of the Czar."

"In that case," I spoke back, "Palestine is not a solution either. You say yourself that the Jewish people will find no security in this world until the hatred for God is eradicated and His Law is revered in the whole world. The Jews will have to wait a long time, I'm afraid!"

"If God's will is to be done by the repetition of words," he came back, "with the recital of credos or with the celebration of ritual ceremonies, surely nothing will come of it. We have been doing that for two thousand years . . . . It has brought us more evil than good; wars, strife, hatred, poverty. It must be done by deeds. In our relations with other peoples we must first of all acquire a respect for the character and instincts and the will and the deeds of others. Peoples cannot love each other. They can respect one another. We must learn that all people—the Jews, too—have a right to work out their own destiny in their own way, according to their own character and talents and national ethos. That is the true democracy towards which we are aiming. Believe me," he ended, "humanity will have advanced a great step on the road to the ideal when the Jews are once more working out human relations in their own spirit. And that they can do only in their own land . . . ."

Time went by and I gave the matter no further thought, although I laid the pastor's words up in my heart so that I still remember them after these thirty years. Nothing more was said about Jews until one day, one of the last days at school, I was once more reminded of the problem. I was then eighteen years old. The class in Hebrew, most of whose members were destined to go to the university to study for the Christian ministry, was reading the 137th Psalm: *"Al-naharoth Babel sham yeshavnu:* By the rivers of Babylon, there we sat down . . . ."

The instructor remarked that the Talmud has an amazing commentary on that Bible text. It says that when the exiled Hebrews answered the Babylonians that they could not sing the Lord's song in a strange land and hung their harps on the willow trees, their captors nevertheless insisted and even threatened them with violence. Rather than obey, the Talmud goes on to say, the Jews bit off their thumbs so that it became impossible for them to play snared instruments. For such touching loyalty to Zion, the cloud of the Shechina, the Divine Presence, overshadowed them, hiding them from their tormentors and wafted them away to a happy land. And no one knows where they were taken.

"I was thinking," said the instructor smiling, "that

Van Paassen, who is leaving us next week to go to America, may perhaps find those Jews on his travels and write us a word as to how they have fared since they became invisible to the Babylonians."

I found them not. I did not look for them. I had other things to do than worry about Jews and their problems. Within a short time after my arrival in North America the first World War broke out, and I was on my way back to Europe as a volunteer in the British army. Even so, in the autumn of 1917 when, upon the recommendation of the War Office, I had been transferred from active duty, because of a shattered left arm, to a branch of the intelligence service where my task consisted of "explaining the meaning of the news" to combat troops, I was again reminded of the conversations back in Holland. I had to tell my comrades one day what a great moral and political act was Britain's issuance of the Balfour Declaration, which promised the Jewish people a home of their own in Palestine.

When the war was over and I had been demobilized in Canada, I entered newspaper work in that country. One of the first pieces I wrote for the *Globe,* of Toronto, dealt with the amazing phenomenon of the revival of the Hebrew language. A local rabbi, Dr. Barnett R. Brickner, had made a trip to the Holy Land, and I was assigned to interview him upon his return. He told me that in the new Jewish community in Palestine the old language was a spoken tongue again. I remembered Renan's words: "A quiver full of steel arrows, a cable with strong coils, a trumpet of brass crashing through the air with two or three sharp notes, such is Hebrew. A language of this kind is not adapted to the expression of scientific results . . . . It is to pour out floods of anger and utter cries of rage against the abuses of the world, calling the four winds of heaven to the assault of the citadels of evil . . . . It will never be put to profane use."

"How then can it serve in a modern community?" I asked Dr. Brickner. "What is the modern Hebrew word for stovepipe, for steamboat, for smoothing iron, for reaper-binder, for electric lamp, for intravenal injection, for brassiere? . . . . "

The Rabbi explained that often the instinctive, instantaneous nomenclature given by children to an unfamiliar object, children who knew no other tongue but Hebrew, of course, was adopted for common use. For instance, a cravat had come to be known by the Hebrew word for herring because a European teacher's tie had thus been designated by his laughing pupils, who had never seen such neckwear before. The mechanical coupling of railway cars was given a name which derived etymologically from the Hebrew word for marriage. I said I, too, should like to hear the language spoken, the language of the prophets, and not sung or drawled as it is done in synagogues, but naturally and simply. And I wrote in my interview that the Jews had put one over on the Italians, the tongue of whose Roman ancestors was dead for all everyday purposes and remained dead.

However, before I was to hear Hebrew spoken in its own milieu, many other things were to happen. I became a foreign correspondent and was stationed in Paris, where I entered the *Ecole Pratique des Hautes*

Etudes at the same time, to study biblical criticism under Guignebert, Lods, Goguel, Couchoud, and Alfred Loisy. One day I received an invitation from the Polish government to tour the country along with a party of journalists from various countries and see the new Poland that was rising from the dust and ashes of centuries. It was in Poland, Bucovina, Volhynia, and Galicia that I saw the Jews, the mass of the Jews, the real Jewish people. The male chaperon the Polish Foreign Office sent out with us was not happy about my suggestion to let us see how the Jews lived. There were so many more interesting things to observe and write about, he said: new industrial centers, new railroads, new mining districts, the new port of Gdynia. Why bother about the Jews who were still adhering to their medieval customs and archaic dress, who were "an alien, unassimilable bloc," not Polish at all, and who had not caught the rhythm of the new order in Poland.

I slipped away from the officially conducted tour and went my own way. I visited the ghettos of Warsaw, Lodz, Lemberg, and Kalisz and the Jewish villages in the depths of the Russian Carpathians and the Polish Ukraine; and I wrote in one dispatch that I had seen human degradation at its most abject.

In Lodz I saw the almost incredible situation of three or four hundred thousand Jews crowded in an area of one square mile. Most of them were weavers of prayer shawls who had been totally ruined by the closing of Russia's religious market. They lived, for the most part, in one-room cottages, two or three families together. In one cottage, where the weaving stool stood in the center, an old man was dying on a mattress in one corner and a child was being born in another. The remaining ten or twelve inhabitants of that room sat in the doorway, although it was bitter cold, so as not to obtrude upon the sacred but here so sordid processes of life and death. Next door in a cellar was a line-up of Polish soldiers extending into the street and awaiting their turn to visit a prostitute whose bed stood behind a curtain while her mother boiled water on a pot-bellied stove for the customers' ablutions.

In Warsaw I climbed the stairs of the Jewish tenements, ten a day for thirty days. I was welcomed into one room where my boy guide's parents lived. There was only one piece of furniture in the room, a bed on which the boy's grandfather lay coughing. The old man scarcely looked up when I entered and immediately turned back to gaze through the cracked and paper-patched windowpanes through which the snow could be seen descending in heavy flakes. My guide's father and three of his brothers sat on the floor. They were sewing buttons on khaki army overcoats. His mother stayed out on the landing, ashamed to show her ragged clothing to a stranger.

When we had sent for some food, we squatted on the floor, and the neighbors came in to share our meal. My guide's mother, too, overcame her shyness at last and entered the room. When we had all eaten, the mother produced a family album from under the bed on which her father-in-law lay. With some pride she showed me its pictures one by one, telling me the names and family relationship of the men, women, and children whose likenesses I inspected: a procession of well-to-do bourgeois Jews, confirmation and wedding groups, grandfathers, grandmothers, men in neat cloth suits such as a bourgeois wears in Holland on a Sunday, men carrying golden watch-chains on their vests, some even wearing top-hats, women in silk dresses wearing bracelets and necklaces, and lovely children, well nourished, round-cheeked, and healthy.

The mother then told me that they, the members of her family and the neighbors who had all trooped in, had not always been as I saw them now, destitute and hungry. Her own husband, for instance, had once owned a big dairy-products store in the main shopping district of Warsaw. And that neighbor, to whom she pointed, a man in a ragged coat and broken shoes, had been a prosperous timber merchant, and still another, who was dressed even more shabbily, had been a cattle dealer with offices in various large cities and with agents all over the countryside.

But now they were all paupers and beggars. And not only those in the room, but all the Jews in the block and in the next block and so on for sixty or seventy city blocks, three or four hundred thousand of them. And it was the same all over Poland, in the cities, in the towns, and in the villages; masses of ruined Jewish middle-class people. Everywhere I saw hollow-chested and pale-faced children in rags wandering about in the narrow, cold ghetto streets. There were no fires in the houses and there was little clothing to wear, although it was the heart of winter. In Kalisz I arrived on a day when the railroad yards were almost taken by storm, as a wildly riotous crowd of Jewish men, women and children clamored around some carloads of spoilt potatoes which had been rejected by the military command as unfit for consumption by the troops. And they, too, those Jews had once, not so long before, been fairly prosperous tradespeople, merchants, restaurant keepers, and professional men.

"Now we are doomed," said my guide's mother. "We have no hope." If her husband and the three boys, she said, worked from morning till night six days a week, they made altogether the equivalent of two dollars a week. "They could probably make a little more," she added, "if they worked after nightfall, but the money to buy oil for the lamp is lacking. And I don't know what is to happen to us when there are no more overcoats to sew buttons on . . . ."

Each man's story was the same: once he had been fairly well to do; now he was ruined. And not only in Warsaw, it was the same in every town and village I visited. Three million Polish Jews were living in squalor and misery that beggared description, without hope of rising one step in the social ladder.

So long as Poland was under the czar, the Jews constituted the class of traders, merchants, and middlemen in that country. With the rebirth of the Polish state, the new masters started building a new middle class composed of the non-Jewish citizenry. The new middle class was built, unavoidably they said (for this was quite frankly admitted in government circles), on the systematic and calculated ruin of the Jewish population. It had to be, the leaders of the nationalist parties told me. There was no alternative. It was a question of life and death for the new Poland to destroy the old middle class and to get rid of it. There were two or three million superfluous Jews in Poland. Ultimately they would have to perish or be evacuated from the country, go elsewhere, anywhere. The Polish authorities did not care where they went, for in Poland there was no room for them.

In Lemberg, a great industrial center, the condition of the hundred thousand Jews was even worse than in Warsaw. There I saw entire families who lived on the ten or twelve dollars a year sent them by Jewish charitable organizations in America. Human beings were so closely packed in the ghetto of Przemysl that the air was foul and unbreathable. I constantly wondered that no epidemics were raging.

As we sat talking in that upstairs room in the slums of Warsaw, I heard the beat of a hammer somewhere in the building. I asked if anyone worked after dark. No, I was told, it was the carpenter downstairs; he was making coffins. "He is the only busy man in the block. He has more work than he can handle. He works night and day, and still there are not enough coffins . . . ."

The mother closed the album. As she did so a slip of paper fell out. I picked it up. It was a snapshot of a young boy, dressed in shorts, bareheaded, his blond curls combed back. He was leaning with one hand against a tree.

"Who is this?" I inquired, as I handed back the photograph.

"That is Mischa, my youngest boy," said the mother.

An audible silence fell upon the people in the room. I saw in amazement that the woman's eyes were filled with tears. The old man on the bed lifted himself on his elbow: "Mischa?" he called out feebly in a questioning voice. The boy must be dead, I thought. I have touched on a delicate subject. I should have been more careful.

"No father," said the woman to the old man, "Mischa is not here. I am telling the strange gentleman about him . . . . " And turning to me she said: "Mischa lives in Eretz Israel (Land of Israel). He walked to Palestine."

"He walked?"

"Yes, he walked. He had no money and no passport. He was notified one day that he must serve in the army. He said he would not serve in the army of a country which denied his people bread and life. He went away and disappeared. Seven months we were in fear and anxiety about him. Then came a letter from Mischa. He had arrived in the Holy Land. He is saving up money to bring us all there."

There was a stillness again. An old Jew wiped his eyes and said to me: "We'll never see Palestine. It isn't necessary. We are lost, all the old people are lost. It doesn't matter, so long as our children may go to the Land . . . . "

"God has promised it to them," I said.

At this everyone burst out in tears, and I, too, wept for Zion . . . .

In a lesser degree but visibly more acute each year that I returned, and matching the increase of tension throughout Europe, the situation of the Jews in Rumania and Hungary and in all that Eastern European hinterland bordering on the Soviet Union, where most of the Jews of the world are concentrated, was as tragic and hopeless as it was in Poland. The

Jews—ninety-nine per cent of them in great poverty—lived from hand to mouth, from day to day, never quite free of the fear of disaster on the morrow, always at the mercy of sudden shifts in political power and of a disturbance of the precarious and already menaced *status quo*. They were surrounded by strong and ever stronger-growing nationalist parties with xenophobian or overtly anti-Semitic tendencies. Everywhere the Jew was looked upon as the stranger, the intruder, the alien, who by his mere presence accentuated the struggle for existence.

The Jew did not look upon himself as a stranger in Hungary or Rumania any more than he looks upon himself as a stranger in America. He had lived in those countries for ages. He had taken root. He was established. He had his civic rights. Some Jews occupied important positions. Some owned newspapers or sat on the bench in the courts of justice. Theoretically the Jew was fully equal before the law with Magyars and Wallachians and Galicians. He was a citizen. Nevertheless his rights were progressively delimited and curtailed. Certain trades and professions were closed to him. Against Jewish students a *numerus clausus* was introduced in the universities and colleges, first tacitly, then openly. Bit by bit he was driven back upon himself by discrimination in industry and governmental services only to be accused then of clannishness, of anational tendencies, of unassimilability. If he dared to point across the border to the Soviet Union, where racial discrimination and the inferior status of second-rate citizenship had been radically expunged, he was accused of Communist sympathies, of harboring subversive, unpatriotic sentiments; and the prejudice against him became more accentuated.

I met Jewish leaders in those countries, brave men and sound democrats, who insisted that the Jews must nevertheless stand firm, not run away, but rather ally themselves with all forward-looking elements in the nations among whom their lot had cast them, and fight it out. But the political allies they acquired that way aways deserted the Jew in the moment of danger or whenever opportunism dictated. When the final showdown came, after Hitler's accession to power, there was none to stand by the Jews, not one. All the liberalism, all the sonorous phrases about democracy and equality of races and good will went for naught; and the Polish and Rumanian equivalents of the National Conference for Christians and Jews evaporated like snow in a summer day's sun. And those who still wanted to see right done—a few revolutionary Christians and a few heroic Socialists—went with the Jews into the concentration camps and into death . . . .

Probably the last Jew to go to the Bessarabian murder camps in Rumania was Dr. W. Filderman, a member of the Rumanian parliament. Dr. Filderman once showed me his home in Bucharest. I marveled at his library and paintings. "See," he exclaimed, "that is the way a Jew lives who is heart and soul a Rumanian at the same time . . . . I am the friend of King Carol, and the Foreign Minister calls me by my first name and always consults me on important matters of foreign policy."

I knew the Foreign Minister, too. He asked me when I entered his office: "Are you a Jew?" When I answered in the negative, he said: "In that case I can shake hands with you. You see, we don't like Jews in Rumania. We are going to get rid of the whole lot of them, every one of them. They either go out abroad, or we'll starve them out . . . . "

I had not the courage to tell Dr. Filderman or the Jewish Senator Nemirover what the Foreign Minister had told me, nor did I tell them what the Minister of War and the proprietors of the big Bucharest dailies had told me.

I went to America for a short vacation at Christmas, 1931. I had just talked with Adolf Hitler in Bonn after one of his propaganda meetings. He upbraided me violently for daring to defend the Jews in his presence. "You, who are an Aryan, a Nordic, a Teuton from the shores of the German Ocean," he screamed, "why do you not see the menace of the Jew to our Western civilization?" I had seen the strange, unearthly fire of hatred in his eyes and was convinced that the man would before long be the master of Germany.

In New York, the day before I sailed back to Europe, I was invited to deliver a lecture in the vestry hall of Temple Emanu-El on Fifth Avenue. I said in the course of my address: "In Germany, Jews have temples as great and magnificent as this. In Germany Jews occupy positions of eminence. There are Jewish mayors of cities, Jewish governors of provinces, Jewish heads of universities. Jews are prominent in art, in culture, in medicine, in the theater, in journalism. Jewish names are inscribed among the benefactors of the German nation on the facades of hospitals, libraries, public baths, and monuments. Yet nothing will remain of it all. Not a Jew will be seen in the Reich in ten years.

Even Jewish names on tombstones will be obliterated . . . . I say so because I have looked in the eyes of Adolf Hitler."

After the meeting I was taken to the home of a prominent American Jew. He met me in the hall. "I heard you speak this evening," he said. "What you said was preposterous. Take it from me, young man, Germany is still Germany. Civilization is not going to be wiped out like that. America and England will also have something to say."

"I hope they will," I said, "but I am not so sure as you are . . . . "

Wherever I went after that, in public gatherings in France, Switzerland, Germany, or Poland; in the Colonial Office in London or at Lady Astor's home in Plymouth, where British foreign policy was often discussed . . . and set; in the Prime Minister's bureau in Paris during Painlevé's, Herriot's, Daladier's, and Blum's tenure of the office; in audiences with Admiral Horthy of Hungary, Pilsudski and Beck of Poland; with Signor Mussolini at the Palazzo Chigi and later at the Palazzo Venezia; in the councils of the Dutch Reformed Church in Holland where I had many relatives and friends, I had but one word on the Jewish question, and I said it "time and untime," as the Dutch phrase has it: the Jews should be evacuated from Europe at no matter what cost. They should be directed towards Palestine in their hundreds of thousands, in their millions, if possible. Just as there had been a total evacuation of the Greeks from Anatolia and of Turks from Grecian Thrace under the supervision of the League of Nations, so Jews should be transferred to their Palestinian homeland by land and sea in the shortest possible time. No matter whether Palestine can absorb them or not; let them be temporarily housed in big detention camps and liberated at a rate warranted by the economic development of the country. But save them, save their lives . . . .

I am afraid I became quite a nuisance by saying this over and over again. Indeed, men accused me of being tormented by an *idée fixe.* Some doubted my sanity. I did not know what I was talking about. How could entire populations be uprooted and transported to a strange land? Wouldn't the Arab world be the first to reject Britain and rise against her if the British Government sponsored such a fantastic scheme? The Jewish community of America, so rich, so influential, so powerful, had certainly not expressed itself in favor of so radical a solution. Who was I then to advocate a scheme that would, if carried into effect, disrupt the normal life of entire nations? Journalistic friends turned their back on me. It was first whispered, then openly written in the Nazi press of Germany, and later taken up in the Dutch government's diplomatic and consular offices in Washington, Montreal, Paris, Munich, Rome, and Cairo that I was of Jewish descent or a converted Jew and that my pockets were stuffed with gold by those Jews who are known as Zionists. Zionists are those people who, in many lands, have set on foot a movement which favors restoring the Jewish people to their own land and helping them to lead normal, healthy lives there. Zionists want the Jews in Palestine protected by those international legal safeguards which protect every nation living within its own boundaries.

There Are No More Prophets!

It is quite true that the rich Jews of America, the upper crust of American Jewish society, "the grand moguls of Judaism," as Israel Zangwill called them once, whose representatives are generally taken by non-Jews as the spokesmen for American Israel, but who are in reality its worse enemies, have little sympathy for the idea of establishing the Jewish masses of Europe in Palestine. I became aware of their attitude on that question early in my career as a newspaperman when my dispatches on the harrowing condition of the Jewish *Lumpenproletariat* in Poland and Rumania remained unpublished in the American journal I represented in Europe. They were held up by a distinguished Jewish leader, now dead, who advised the editors that my reports were exaggerated, alarmist, unobjective, and not in accordance with the information he received regularly from his informants over there.

Nor, it was objected, could that poor, insignificant little country of Palestine in Asia Minor about which I wrote offer a solution for European Jewry's woes. My solution, which was not my solution at all, but the solution advocated by the intellectual and religious elite of mankind, was a nationalistic one, and they, the prominent Jews of America, should not be thought to favor such a crude and untimely venture. The whole trend of history was away from nationalism altogether. They had nothing to do with any foreign nationalistic undertaking. Nationalism bred strife. They were for peace and amity among nations. They were Americans, one hundred per cent, perhaps a few per cent more, *plus royaliste que le roi*. They were Americans of the Jewish faith, a religious denomination, nothing more.

Making a high virtue of a cruel historical fatality, they proclaimed Israel's mission to be dispersal among the nations of the world, to be a light unto the Gentiles and an example to the peoples. They said they believed in the tradition of justice which they considered their heritage as American citizens of the Jewish faith. "What the Jewish people want more than a home of their own," said one of their spokesmen, "is the right to call any place home."

Like that clever cuckoo, I suppose, which lays its eggs surreptitiously in other birds' nests?

Any place! Germany, for instance! Or Poland, whose government in exile in London still talks of evacuating its surplus Jewish population after the war. Or Santo Domingo, or the jungles of British Guiana, or the desert of Libya! Or Brazil or the Argentine, any place on earth, so long as the European Jewish masses are kept out of sight under a thin blanket of charity. No matter if new minority problems are created in Bogota or Peru, to be followed by new friction, strife, new massacres, and another exodus in twenty years' time. Anywhere, so long as it isn't Palestine or . . . the United States of America.

For of America as a haven of refuge for the Jewish masses, these Americans of the Jewish faith dared not speak either. They knew full well that if they proposed mass immigration of Jews into the U.S.A., they would rouse a storm of protest and indignation in this country. They are kept busy as it is, explaining and defending their own position in the United States, citing their contributions to American culture, disassociating themselves from Jewish nationalism, communal autonomy, and all other movements and institutions to which other groups turn or may turn spontaneously without fear of arousing suspicion. In their hearts they know that they are not accepted as Baptists or Roman Catholics or Lutherans or as any other religious denomination such as they claim to constitute.

They are not aware of the fact that their nervous fear of life, the fear of their own people, their self-hatred and self-abasement and servilism which find expression in that multiplicity of peculiar defense mechanisms and organizations they have set up in America such as no other religious or national group has felt called upon to do, and, above all, the bland denial of the existence of one of the components of their distinctive character are the surest symptoms of the Jewish people's mortal malady: the lack of a homeland, the lack of a national backbone.

Humanity sympathizes with a strenuous aspiration. It cannot have respect for people who lack self-respect. The Gentile perfectly understands the Jew who owns up to it in a forthright manner that he is the son of Israel, with its great past and traditions and all its faults and shortcomings. He sees before his eyes—for to the Gentile, even to Adolf Hitler Judaism is one and indivisible—the real, tangible link between the American Jew and the suffering Jewish masses in Europe on the one hand and the reborn Jewish community in Palestine on the other. There lie the Jewish people's roots, and that land with its reborn Jewish community may again be the spiritual center of the Jewries of the world, even as the spiritual center of Catholics is Rome.

The Gentile knows that behind the religious community of the modern Jew lies a territorial past and that Israel is not merely a religious community of great antiquity but the broken and shattered remnant of a people with a soil, dynasties, wars, and a magnificent epical literature that has molded the thought and the art of half the world. He knows that it is neither wise nor legitimate to deny a people's past or to see a nation as unrelated and detached from its past. A living people is but one page in its book of life.

What the Gentile does not understand is the Jew who pretends that he is, except for a slight credal difference, a circumcized Unitarian and that the matter goes no deeper or further than an aquiline nose or slightly bulging eyelids. The Gentile knows that the Jew belongs to an Oriental people who have lived two thousand years in the Occident, but that he retains that which is his greatest glory, the imprint of the East, his ethnical, racial, and national characteristics and that to a certain extent he is, happily, not assimilable to the crude and darker aspects of Christian Western civilization, with its glorification of violence and lack of mercy. Just because the Nazis have made a gruesome caricature and a mockery of race and blood, we do not need to be so childish as to repudiate or hide our origin and our identity.

The Jew should not assimilate and should not be asked merely to assimilate, for the grandeur and desirability of democracy resides precisely in differences and in a respect for the differences in contribution that peoples, individuals, and groups may make towards the sum total of civilization and culture in their own way, according to their own peculiar gifts and *Eigenart*. The reverse is the totalitarian monstrosity of *Gleichschaltung,* the dehumanization and depersonalization of man.

Only the Zionist Jew is a true American patriot, for that Jew would be the transmitter of the Hebraic values which his nationally reconstituted people in Palestine work out and which he would adapt to the enrichment of American life. Palestine is not the American Jews' national home. It is the national home of the Palestinian Jews and of the homeless Jewish masses in Europe. To the American Jew Palestine can be a spiritual home, his "hill of inspiration," the place where his own Hebrew spirit again "bloweth where it listeth," whence the hymn of his ancestors comes over to him and where the performance of his modern kinsmen inspires him with hope and renews his youth and courage in his own environment. The new Palestine is the voice of Jerusalem which speaks again to all, but chiefly to and for its own children wherever they are. Only as a nation, living, working as a nation can the Jews be an example to other nations. Zionism has nothing to do with nationalism in the sense of chauvinist imperialism. It does not seek to grow strong at the expense of others. It is self-determination and auto-emancipation. I hope Palestine will never have to become the American Jews' national home as it became the national home for the once so skeptical German Jewry, which faced less anti-Semitic opposition in 1930 in the Reich than American Jewry now faces in the United States.

I believe that one of the greatest disasters that has

befallen the Western world is the elimination of Israel in a national sense from participation in the work which is the aim and object of all man's strivings: to lift conditions and human society gradually and constantly more in the direction of the Kingdom of God on earth. I believe also, therefore, that Israel in his own home, permeated with the reality of his own Torah, carries the key to his own happiness and to his part of the solution, which is the highest and final solution, that of the Prophets and of the Christian Fathers, from Saint Augustine to John Calvin and Karl Barth: *le thaken olam be malkut Shaddai,* the transformation of the world into the Kingdom of God.

We Christians failed to recognize the significance of it when Hitler singled out the Jews for his first and most brutal attacks. We did not see that the Nazis moved with sure and unerring instinct when they began their assault on civilization with the Jews. It could not have been and cannot be otherwise. We did not realize that what Hitler attacked in the Jews were our own holiest possessions. He recognized in the Jew the bearer, the unconscious bearer perhaps and often an unwilling bearer, of a concept of civilization which places justice in the foreground as the cardinal principle in the relationships between man and his fellow. He attacked the Jew because he saw in the Jew an advance sentinel, the man standing on the outer bastion of that complex civilization which, in spite of many grievous shortcomings, is nevertheless stamped with the Christian imprint and bears the Christian name.

The Jews were the predestined enemies of Nazism

not because of anything they did do or did not do. A people which has its life and being in two concepts—unity in the law of this world and a hope in the final triumph of justice—is stamped *a priori* and ineluctibly as the natural opponent of the pagan idolatry of the state and of totalitarianism. To make Europe *Judenrein* —to eliminate the Jew—was therefore to Hitler and the Nazi ideologists like clearing the way for the final assault on those ultimate values: human brotherhood, collaboration by the free moral consent of the peoples, and democracy, which are Christianity's heritage from Judaism and its essence.

Today the responsibility for all the evil of our time is heaped on the head of Adolf Hitler. "One man and one man alone," it is said, plunged the world into chaos and caused humanity to wade into a river of blood and tears. One man and one people, which followed him, brought on the catastrophe. In this view of modern history we ourselves are without sin, without a share of responsibility in the matter. Cordell Hull expressed surprise that the dictators could "overnight almost" sweep along tens of millions in Germany and Italy. It started all of a sudden with the March on Rome, with the seizure of Manchuria, with Herman Goering setting the Reichstag on fire. From those incidents, which by themselves were perhaps not so phenomenal date all our sorrows, all our pain.

But who helped set Mussolini on the throne and who placed Hitler in command? Evildoers helped them, men without God or conscience? Was evil then so all-powerful before the present deluge that it could not be resisted? Where were the others, the good and the decent, who could still distinguish between truth and lies? Did the representatives of decency, of intelligence, of justice, of religion at least speak out? Why were they silent in all languages? Why did they abdicate? Why did they only begin to take note of rise of evil when it touched their own lives? Didn't Churchill call Mussolini a genius whom he would gladly follow if he were an Italian? Did not the Catholic bishops of Germany, in Fulda assembled, three years in succession praise Hitler as a man of God? Didn't the Pope call Mussolini "providential" and did not the Vatican send its highest distinction but recently, in July, 1943, to Ion Antonescu, the black murderer of a hundred thousand of Rumania's defenseless poor?

The disaster did not come overnight. It was preceded by many portents and warnings. It is a long story. There were many tryouts of horror, dress rehearsals, first-night presentations as it were, when the critics went to see the show and have their say—improvisations of horror that preceded the day when horror itself appeared on the stage in all its full, shameless frankness. For a quarter century and more the tryouts have been succeeding each other with regular sequence: the willful starvation of millions by the *cordon sanitaire* thrown around Russia, the embargo on food to the Weimar Republic, the extermination of Greek Orthodoxy in the provinces of the Ukraine captured by Poland, the bombing of villages in India, the soaking in poison gas of Ethiopian peasants, the massacres in Spain, the besmirching of the Popular Front in France. Tryouts they were, rehearsals of horror by Christian nations with the critics looking on, the guardians of the conscience of humanity applauding and the representatives of the Church sometimes giving the go-ahead sign with a nod of assent and a blessing.

The rehearsals were a success. They were eminently satisfactory. God was thanked for having them brought to a splendid end. And what then? Then the real show came, of course. Then, inevitably, quite naturally came the hour when the beast of Apocalypse, which steals the crowns from the brow of humanity and piles them on its own forehead, the brute without conscience or compunction which tears up the Gospel generation after generation, mounted the world's stage for the final triumph. Then the Beast, eternal Ashur, could go on stealing and calumniating and massacring without break or check. Then it went beyond the deeds of Nero and Attila, befouling human nature, growing drunk on the blood of the saints, and no longer assassinating human dignity, but tearing it to pieces and finishing it off in the slime of its own saliva.

No, it did not start with the March on Rome, with the invasion of Manchuria, or with the Reichstag fire, but much earlier. It started at that point in history when Christianity with its faith and love withdrew from the world and created itself a personal sphere, a sacrosanct retreat by confining the operation of Christian virtues to private circumstances and affairs.

It is by no manner of means a coincidence that today's triumph of evil was preceded by thirty years in which abdication and withdrawal from the world were openly preached and boldly characterized in our Protestant churches, on the model of Germany, as *"die Selbstbesinnung der Kirche auf ihre eigene Sache,"* the Church's concern with and concentration upon her own business. Yes, we went so far in our ecclesiastical pride in our Protestant churches to deride profane humanism, unaware of the fact that the world-conquering divine thought which came on earth with Christ, could not otherwise manifest itself save in extraecclesiastical and even nonreligious forms. Secular enlightenment had to do what the Church neglected to do: to champion human rights, equality before the law, freedom, the right of the Jew to live, when the Church seemed unaware, or deliberately ignored the fact, that these demands issue logically and consistently from her own teachings.

I heard an uncle of mine preach one day in St. Bavo's Church in Haarlem on the text: "The Kingdom of God comes like a thief in the night." "Most people," he said, "are so romantically minded that they think the thief will make a noise and wake them up. But he will not do anything of the kind, if he can help it. The occupants of the house will sleep right through his nocturnal visit. So it is with the Kingdom of God. It comes and we know it not. In the morning we do not recognize it and its features. Often and this is especially true of our time, we do not realize what are the demands the Kingdom makes upon us."

That abdication before the world, that flight from reality, became also characteristic of other, nonchurchly spheres of the spiritual life. Science, art, and literature can point to few luminous exceptions who were not silent. And this does not apply only to certain countries, to Germany or Italy for instance, for the frightening experience of the last few years points precisely to the spiritual supineness, the universal lack of moral assistance and backbonelessness throughout the whole world.

Nowhere does there seem to have been sufficient moral strength and clarity of vision to appreciate the true nature of the powers of chaos. On the contrary, there was again and again a diligent search for a *modus vivendi* with these powers. All sorts of excuses and justifications for their appearance and growth were discovered and advanced. They were approached for possible collaboration. They were offered compromise after compromise. They were offered one sacrifice after another.

In this tragedy was revealed the inner breakdown of a world which still confesses itself Christian or at least respectful of the ideals which stem from Christianity. It is not a question of modern society having given up its faith and ideals. It is a question of the inner paralysis of society by a secret poison.

That poison has its seat and origin in our own unacknowledged guilt, our own failure to assume responsibility for the rise of evil. It is this guilt whose secret poison influences our political action in that it disarms us morally and renders us obtuse and impotent in the presence of political injustice and crime.

Iam non est Propheta! There is not more any Prophet! By a prophet I mean a courageous speaker of the truth, a man who by virtue of a higher, divine authority dares to tell the mighty of the earth in concrete cases and in specific circumstances: "Thou Shalt!" or "Non licet! . . . "

If the Church does not speak the prophetic word in our day, but instead stands powerless, fearful, and silent in the presence of truly apocalyptic events or at best mumbles inane and innocuous commonplaces and generalities in that jargon of traditional piety which has become unintelligible to the men of our generation, it is because she thinks too much of her inner organization, of the abstract eternal verities of which she is the depositary, of her temporal power, of the good repute she enjoys amongst the well-to-do whose chief concern is their own tranquillity of mind and the undisturbed enjoyment of their comfortable position in society. It is with the authority of this "good enough" armor of respectability, smugness, unction, and sentimentality that the Church thinks to confront a world in turmoil.

But the potential strength of the Church always lay and still lies solely in the old saying: "I have spoken and (therewith) I have saved my soul," in other words, in the freely received and freely confessed or spoken word of truth. Only by fulfilling her duty of confession, of speaking out at no matter what cost or risk in the language of our times, to the men of our time and on subjects and problems which directly and intimately concern the men of our time, can the Church still save her soul and perhaps the world. For that, however, she must stand forth to give battle and fight and struggle, with a courage and frenzy born of desperation, *jusqu'au sang,* until blood flows if needs be, for the day is far spent and from the enemy's camp comes the swelling sound of the triumph's song.

Yet what do we see? Precisely in our day when there have been unleashed, and this by no means in Germany alone, demoniacal powers which seek to depersonalize the individual, dehumanize man and ultimately bestialize him, when man is no longer a free agent, the arbiter of his own fate but more and more subject to intellectual and material intimidation, when Western civilization is in full process of decomposition and a spiritual debasement without parallel spreads through the world like some black cloud of poison gas, the Church still shows no symptoms of alarm or of disquietude, still goes on feeling fat and at ease in Zion. She enjoys official respect and recognition. She is protected and petted by the State. Churches and ecclesiastics play *il gioco del mundo,* the game of the world, as Pope Julius II once called the Machiavellian game of power politics. Religious instruction is handed out in abundance. It has become almost obligatory. Freethinkers are silenced. Teachers are harassed; men of independent mind and research are driven from the institutions of learning. Books and pamphlets are appearing praising the Inquisition as a useful institution for our time. The police is brought in to do what the spirit cannot accomplish—perhaps because the spirit is nonexistent . . . .

Instead of a new form of Christianity emerging from the friction and contest of elementary forces and from the new generative processes at work in society, a new form of Christianity that would give the answer to the distress of humanity in that the central place would at last be given to the luminous message of the Kingdom of God on earth, the old egotistical religion is coming to the fore again, that ancient and efficacious "opium of the people" which clamps itself, almost frantically, and for a good reason, to an affirmation of individual salvation and a belief in the hereafter—a creed that accords perfectly with the schemes of the powers of evil.

Sometimes it seems indeed as if the days of Constantine the "Great," the Church's unholiest hour, when she was made *hoffaehig* and quality was sacrificed to quantity, have returned in a new form. As then, when that sovereign, who had come to power by all manner of ruses, lies, and violence, thought it opportune to adopt the symbol of the cross in order to further his ambitions, in the same manner act today the big demagogues of a similar type. And even as the vast majority of the official and semiofficial representatives of the cause of Christ broke into paeans of joy over the Emperor's "marvelous conversion by the grace of God," and a small number only were aware that a frightful catastrophe had occurred, so it is again today.

A catastrophe! For never does the cause of Christ suffer as much as when state and society go out of their way to fawn upon the Church and praise her; reversely: never does that cause prosper so well as in the hour of persecution. One would almost feel like congratulating the Church if she could again become the object of harsh and bitter persecution, if she were driven out of her haughty basilicas, marble churches, and oak-panneled vestries and be forced to crawl back underground into the catacombs; if the servants of the cause of Christ were denounced once again as Beelzebubs, as they were in the days of Jesus, that is, in modern language, as Bolsheviks, reds, traitors to the state, as fools, bandits, fanatics, false prophets and seducers of the youth; if they were reduced to the last extremity and menaced with dire things, with death itself . . . .

Alas, it is not so. The Church is prosperous. The Church is at peace with everybody. She loves both sides equally, good and evil alike. She is at peace with all men. She is neutral . . . .

But a straightforward conscience cannot live at peace with all men. It cannot and must not. If a man discovers that he is at peace with the whole world, he had better examine himself and ask to which of his obligations he has been unfaithful. A free conscience is duty-bound to rise against the declared enemies of the Divine, which it should instinctively recognize. That is the prophetic intuition. It must rise against obscurantism, against all violations and abuses of power, against all dogmatisms, against the vilifiers of the human soul and the monopolists of privilege. "No one has the right to be neutral in moral questions," wrote Alfred Loisy to Pope Benedict XV in the course of the First World War. "Whoever pretends to be neutral in matters where justice is concerned fails to be impartial. As a matter of fact, whoever pretends to be indifferent is in reality siding with him who is in the wrong and against him who is right."

There is nothing, literally nothing, no cause, no hope, no virtue, no threat of catastrophe even, that really touches people's hearts in our time. Nothing captivates them, neither heaven nor hell. Nothing is capable of shaking them, of setting them on fire. If anyone wants to go to battle for this or that cause or question or problem, the men of our time immediately draw the conclusion that he does it in order to profit by it, or to create a sensation, or to satisfy some secret ambition, or to draw attention to himself.

When anyone uncovers abuses, combats injustice, castigates meanness, the slavish souls ask at once: what has he got in mind now? What is he driving at? What objective has he in view? They soon gather around the village pump or in the newspaper to snicker and whisper that he does it to have his mouth stopped with gold, that his motives are selfish and contemptible, that he does it out of vanity or just to amuse himself and make sport. In the presence of any act of heroism or courage or moral greatness or grandiose madness these twaddlers, barbers, and empty singers of an idle day ask: why is such a thing done?

Don Miguel de Unamuno, the Spanish mystic, said once in Paris in my hearing: "When I make some proposition to a man these days, when I tell him what in my opinion should urgently be done, I always hear the question: 'And after that?' To such questions I answer with another question. When anyone says: 'And what afterwards?' I say: 'And what was it before'?

"There is no such thing as the future," said Don Miguel. "There never was any such thing as the future. What is called the future is a lie. The real future is today. What will become of us tomorrow? There is no tomorrow! What is happening to us today, that is the only thing of importance."

The importance of today! Men are not only *not* worried about today, they are entirely satisfied with it, with the bare, naked existence of today. And there lies the pernicious evil of our time. That there must lie something beyond today and beyond everyday existence does not trouble them. It leaves them unconcerned. They have no nostalgia for tomorrow, for they do not act today as if they would be prepared for the today of tomorrow. They are like the members of that tribe Stanley encountered in the heart of Africa and to whom he explained that beyond their impenetrably dark forest lay fields and rivers and mountains and sunshine. They shook their heads and moved their arms in a gesture of unbelief and hopelessness as if to say: no, all the world is like this, obscure, terrible, dark, unchanging. They refused to follow Stanley into the light. There was nothing desirable or hopeful beyond their somber, pitiful existence.

But did they really exist? Do people really exist who dare or will not look at today? Does a Church really exist that does not tremble for today? I do not believe it. For if she existed, really existed, she would suffer under the agony of today's existence and not be satisfied with it. She would question the sacrosanctity of today, of the *status quo,* grow hot with indignation or cold with terror and apprehension, and rebel against it. If the Church really lived in time and space, she would suffer under it for not living in the eternal and in the infinite.

A Chinese philosopher said to André Gide: "The amazing thing about Western civilization is that it bears the name of Christian and acts as if mercy, love, justice, and brotherliness were the cardinal evils to be avoided. Why do you not do one thing or the other: drop the Christian name and stop insulting the name of Christ by frankly pursuing a materialist objective or follow Christ practically and take the consequences?"

Take the consequences! Ah, there's the rub! That we dare not. We Christians stare ourselves blind on our abstract "eternal truths" and believe we have done our duty nobly towards society when we keep them unsoiled and defend them against attacks. We are more concerned with the teaching in itself than with the soul that is to be taught. But the question is precisely to know not what is eternally true in itself, but how to apply that eternal truth to our time, here and now, in definite and specific human relationships and circumstances. To know when to apply the eternal verities and when to proclaim them: that is spiritual leadership. To that end serves the prophetic office! To what other end, pray? "Whoever," harshly remarked Emil Brunner, one of the few remaining prophets of our time," whoever says that he has the eternal Word and does not translate it into the language of his time and adapt it to the exigencies of his time does not have the eternal Word at all. He is a liar."

If we fail to apply the divine truths to concrete historical circumstances, in the reconstruction of society, in the building of a new order, we leave the road wide open to the Devil's powers. To keep the truth hovering in the abstract will lead to a passive acceptance of all manner of evil conditions and in the end to a concordat with the Devil himself, who is quite willing to respect our churchly formulas, liturgical sonorities, and pietistic speculations about the inner life and the life hereafter, as long as he is left to rule the present. Today is what counts.

The heaviest burden to bear is the suffering of the innocent, the thought that we with our own hands and minds, by deeds of commission and of omission, and by our silence often, helped to increase unnecessary suffering in a world in which there are enough painful riddles as it is. In this respect we are all guilty from the highest to the lowest, all those who call themselves Christians, also he who writes these lines. For he does not presume in writing of these things that he is "the one just man from whose hands the conduct of the universe has been inexplicably abstracted." He is the more at fault because he saw more than many others. He, too, was often silent when he should have spoken—no, when he should perhaps have cried out no matter how shrill his voice and how grotesque his behavior . . . .

I am convinced that Hitler neither could nor would have done to the Jewish people what he has done—perpetrated the most heinous and gruesome crime of history on a helpless, disarmed, national and religious minority—if we had not psychologically and actively prepared the way for him by our own unfriendly attitude to the Jews, by our selfishness and by the anti-Semitic teachings in our churches and schools. For there lies, in my estimation, the crux of the matter, the deepest source of hostility towards the Jews: in the system of dogmas superimposed upon and interwoven with the simple story of the Gospel. Hitler could not have killed the Jews of Europe unless the Church of Christ had first killed them with its fiendish anti-Semitic teachings.

"It is hypocrisy of the lowest and most obscene kind," says Kenneth Leslie,* "for any Christian Church or

---

*Chairman of the National Commission to Eliminate Anti-Semitism from Textbooks, an organization with a membership of two thousand Christian pastors and educators, 521 Fifth Avenue, New York.

group of churches to pass resolutions or to utter words of pity for the Jews before renouncing openly and abjuring forever the horrible reversal of the teachings of Jesus without at the same time demanding that our government open its doors to the martyred Jewish masses and without opening the doors of their churches and their homes to Jews."

Even today the German people, our fellow-Christians in the Reich have not heard a word of warning from our churches. The governments of the U.S.A. and Great Britain have weakly protested, it is true, but not to Germany. They have not acted as if they were really concerned at all about the Jewish tragedy. After the war, Mr. Roosevelt and Mr. Churchill have told the Jews, everything will come right. In the meantime Hitler keeps on murdering Jews, thousands each day, without relenting, even when the signs tell him that he must inevitably be defeated.

If the Führer should tomorrow announce that he is going to make Europe *Bulgarenrein,* as he once promised to make Europe *Judenrein,* and if he began carrying the blood purge of the Bulgarians into effect, and if the governments of Britain and the U.S.A. warned him that for every Bulgarian slain they would exact ten German lives, the Nazis, I am sure, would desist at once. The Bulgars are not our allies, but the Jews are. In the interest of saving Jewish lives, no such warning has ever been given to Germany.

Wherever thou seest the footprint of a man, there God has gone before thee. These words were written in a time and in a place when man's bare feet left their mark in the soft, rain-soaked soil or in the still sand. On the asphalt of our great cities, in the factories, in the rice plantations of the East, on the cobblestones of the ghetto man's feet leave no imprint today. For that reason we cannot leave off looking for him. He must still be sought. He must nevertheless be brought home. Our search should become more intensive as the circumstances grow more difficult. Our sense of responsibility should grow in intensity as the light of day becomes fainter. It is dark in the world. But it is not so dark that the face of a brother cannot be recognized. To find the imprint of human feet in a time of dehumanization is the only proof we can have that we are on the road from nihilism to life, from death to the kingdom of God.

The Balfour Declaration, issued by Great Britain in the midst of the last war and ratified by the Government of the United States of America and by fifty other states after the war pledged the establishment of a national home for the Jewish people in Palestine. It was the cup of grace and communion offered Christendom, the means to redeem itself from the crying injustices and violence it imposed upon Israel in the course of his long and sad sojourn in our midst.

What has become of that project which was to have been the chief instrument to solve the Jewish problem after the last war?

Where has it led? How has it affected the fortunes of the British Empire in the past; how will it affect them in the future?

What has been and is the reaction of the Arab peoples who were affected by the settlement of a new

Jewish community in the vicinity of their immense but empty empires?

The treatment meted out to the Jews by any given country has often been called the measuring rod of that country's civilization. How do the United Nations propose to treat the Jews who escape the holocaust of war? How do they propose to solve the unsolved Jewish question today, or when hostilities cease?

Is the manner and spirit in which the Jewish question is approached today by the Great Powers to be indicative of the way in which the still greater problems of other peoples in Europe and Asia are tackled?

How have we come to the present state of affairs where the Jewish people are being exterminated on the one hand and their means of escape, on the other hand, is cut off?

It is a long story which goes back to the time when the last czar of Russia still sat on the throne and the Great Powers were waiting for the death of the Sick Man of Europe in order to divide his domains in the Near and Middle East.

CHAPTER 2

PRELUDE TO PALESTINE'S LIBERATION

ONE of the most complicated diplomatic maneuvers of modern times was the attempt by Czar Nicholas II of Russia and his advisers to pick a quarrel with the Ottoman Empire in the late summer months of 1914. Contrary to what has generally been said on the subject of Turkey's entry into the war, the state archives of Russia reveal that the Turks were not in league with the Germans "from the very beginning." They were lured into the war by an extraordinary play of power politics on the part of the Russian imperial government. It is this imperialist chess play which had a direct bearing on the fate of Palestine and of Arabia and which still determines Britain's role in the Near and Middle East.

The ruler of Turkey, Mohammed V, had no desire to fight, especially against so powerful a combination as that presented by Russia and her British and French allies. For a generation or more the Sultan, or his empire, had been known as the "Sick Man of Europe" who was now, in 1914, by the general consensus in the chancellories and on the stock exchanges agonizing in his last extremities. He had been badly beaten in the last two wars he waged. On top of that Turkish finances were in grave and growing disorder, while the Sultan's spiritual hold on the Moslems of the worlds as Khalif-ul Islam, or Commander of the Faithful, was fast slipping from his hands.

What could be a wiser policy under these circumstances but to declare for a strict neutrality in an armed conflict in which five great empires, none of them a particularly good friend of Turkey, were already at each others' throats?

Upon Petrograd's urgent inquiries, therefore, as to Turkey's intentions, which had been made known to the world as those of a benevolent neutral to everybody concerned in the war, the Sublime Porte replied that, as a further token of Turkish good will, all troops stationed on the Caucasian border would be withdrawn to inland positions in Anatolia. This would, it was "obediently and humbly" pointed out by the Turkish Foreign Office, enable Russia to concentrate all her military effectives on the German frontier where she had but recently won a signal success with the occupation of East Prussia.

While Nicholas II was making these critical representations at the Turkish Foreign Office, he was secretly instructing his ambassador in Constantinople to do all in his power to favor the consolidation and expansion of German interests in the Ottoman Empire "in view of the fact that the existing Turco-German commercial pacts offered as yet the only visible pretext to break off relations with Turkey." In one dispatch the Czar upbraided Turkey for allowing German interests to install themselves in the Ottoman domains; in another secret diplomatic note, sent almost simultaneously, he urged his ambassador to drive Turkey even deeper into the arms of Germany. The most desirable development would be, he hinted strongly to his ambassador, for Turkey and Germany to conclude a military alliance that would be directed against Russia.

The Turks did not fall into that ambush. In order to appease His Russian Majesty's aggressive diplomacy still further, they offered to open the Dardanelles to Russia's Black Sea Fleet. When the Czar still was not satisfied and secretly reiterated his instructions to his ambassador not to leave a stone unturned to drive Turkey into an impasse from which the only issue could be a recourse to arms, the Sultan offered nothing less than a military alliance and the speedy dispatch of important Turkish military forces to the support of Russia in her war with the German Reich.

The Turks could not get it into their heads, or rather they pretended to be so naive as not to understand that the Czar did not want them as allies, no matter how "humble and obedient," but that he wanted them as enemies. What is more: Russia was growing impatient about the matter. The Czar wanted his war with Turkey quickly, without delay. That poor man, his ministers, and his chief military advisers, all of whom were to be swept into the shades of oblivion by the storm they raised, were still under the illusion in the fall of 1914 that the war with Germany would quickly come to an end through "circumstances independent of military operations." The Franco-British-Russian allies had indeed solemnly agreed not to enter into separate negotiations with Germany and to fight on until the

Kaiser agreed to unconditional surrender, but in the first week of September, 1914, the Petrograd wireless station was intercepting messages flying between London, Washington, and Paris which, upon being decoded, revealed that Britain was cautiously sounding the American Secretary of State, William Jennings Bryan, about mediating between herself and Germany.*

Upon learning of these peace feelers, the Czar, who stood victorious at Königsberg, more than halfway between the Russo-German frontier and Berlin, made it known in London and Paris that before an armistice should halt hostilities on the Western Front and he, not being able to withstand Germany alone, should be forced to follow suit in making peace with the Reich, the prize which had haunted the imagination of the Pan-Slavs for generations must be his. He did not want the Turkish army as an ally, nor did he covet the Caucasian provinces of the Sultan. He wanted Constantinople, the city known in the Russian language as Czargrad, the City of the Czar. Britain and France could do what they liked, but Nicholas was not going to demobilize until he had realized the ambition of all his imperial ancestors . . . .

"Constantinople must be ours," wrote Dostoevsky in his journal as far back as 1899. "It is we Russians who must take it from the Turks, and it must remain ours in all eternity . . . . To us alone it must belong, for what is the Oriental question but that? Fundamentally, the Oriental question is the future of Orthodoxy. The

*Both Count von Bernstorff, German ambassador in Washington, and M. Bakhmetieff, Russia's ambassador to the U.S.A. mention the British secret *démarche* with the Secretary of State and President Wilson.

future of Orthodoxy is inseparably bound up with Russia's vocation and destiny . . . . It is our gospel . . . . The Oriental question has not been invented, as some say, by the Slavophiles or by anyone else. It came into the world by itself, long before the Slavophiles existed, before any of us were born, before Peter the Great and the Russian Empire . . . . The Oriental question is the eternal idea of the czardom of Moscow . . . . Constantinople is the key to our own house . . . . "

The Russian statesmen and the intellectuals who fabricated the ideology of empire presented their aggressive project against Turkey as a crusade, but one may as well call Napoleon's conquest of Palestine a crusade or see in *The Brothers Karamazov* a symbol of the Trinity. The simple truth was that the new and growing Russian industry had to force a way, by war if necessary, to the markets of the Near East and the Mediterranean basin. The czars had fought three or four wars with Turkey to that end. Their last great effort, when it was about to succeed, had been frustrated by Britain and France at Balaclava and Sebastopol.

In the autumn of 1913, the Russian Foreign Minister, Sergei Sazonov, submitted to Czar Nicholas a memorandum wherein he expressed his doubts on the solidity and longevity of the Ottoman Empire. "All the Great Powers without exception," he wrote, "envisage an early dismemberment of Turkey and are even now occupied in consolidating the bases of their political pretensions in the future division of Asia Minor . . . .

"The closure of the Straits by Turkey," he wrote, "has had a nefarious influence on the economic life of our country and has once again shown us the capital importance of the Oriental question. If the complications in Turkey today mean the loss of hundreds of millions of rubles to Russia, what would happen if, instead of belonging to Turkey, the Dardanelles should pass into the hands of a power really capable of opposing itself effectively to Russian demands (to keep the Straits open). And for this it is not even necessary that the Power controlling the Dardanelles be a Great Power. Such a Power would inevitably become a Great Power by reason of the exceptional geographic conditions. Indeed, he who controls the Straits holds not only the keys to the Black Sea and the Mediterranean, but also the bases for an expansionist movement in Asia Minor, the Near East, and Africa and will in addition be assured of hegemony in the Balkans. In other words, the power which replaces Turkey becomes the heir of Turkey and will follow the road the Turks have made themselves to the Red Sea and the Indian Ocean .... We cannot envisage the possibility, either morally, economically, or politically, of a foreign power installing itself on the Straits .... "

In the light of Sazonov's memorandum the speed which British diplomacy deployed in trying to bring the war with Germany to a quick termination appeared to the Czar in an entirely new perspective. If that war ended now, in September, 1914, Russia would also be compelled to sue for peace and would emerge without having settled the question of Constantinople. If the war ended without Turkey's participation, how could Russia obtain the key to her own house? What guarantee had Russia that Britain and France, possibly with the collaboration of Germany, would not subsequently proceed to the dismemberment of Turkey without asking Russia to share? Hence the war must be made to go on, and Turkey must be brought in as quickly as possible.

A new fortuitous pretext to break off relations with Turkey had been injected into the situation in the meantime: two German cruisers, the *Goeben* and the *Breslau,* had entered the Dardanelles and had not been disarmed by the Turks. In alarm, the Russian Prime Minister instructed his ambassador in Constantinople not to protest, for a protest by Russia might lead the Turks to disarm the German battleships at once and intern their crews. The Turks were too apt to give Russia complete satisfaction on this point, as they had done on every occasion when some issue between them and the Russians had arisen.

This time, however, Enver Pasha, the chief of the Turkish government, fell into the trap. Whether it was by sheer negligence that he failed to disarm the two German cruisers, or whether the German diplomatic mission in Constantinople persuaded him to let events take their course, cannot now be said with certainty.

In Winston Churchill's version of the matter, the Turks had been on the warpath since the month of August, 1914, when, he says, they had concluded a secret military accord with Germany and thereafter merely looked for a suitable pretext to enter the war against the Entente. But Mr. Churchill also states that Russia's position *vis-a-vis* Turkey was "eminently correct" at the outbreak of the war in that she had made a joint declaration with France and Britain that the territorial integrity of the Ottoman Empire would be respected at the peace. Sazonov's memorandum to Nicholas II and the published memoirs of the French, British, and Japanese ambassadors to Russia, make it undubitably clear that one of imperial Russia's chief considerations in bringing about the upheaval of 1914 was the dismemberment of the tottering Ottoman Empire, the seizure of the Dardanelles (the key to her own house), and commercial expansion in the Levant and the Near East. Not a single reference to a secret Turco-German treaty dating from August 2, 1914, could be found in the archives of the Russian Foreign Office by Leon Trotsky, who made a diligent search when he assumed the position of Foreign Commissar in 1917.

On September 1, 1914, having learned that the Turks had not disarmed the *Goeben* and the *Breslau,* M. Sazonov heaved a sigh of relief and said, according to a *procès-verbal* of the cabinet meeting, to the Czar and his advisers assembled at the imperial summer residence of Tsarskoe Selo: "At least a Turco-Russian alliance is no longer possible. And soon we shall see more!"*

How little sincerity the Czar's government attached to that joint declaration with Britain and France at the outbreak of the war to respect Turkey's territorial integrity became quite evident on September 10, when the Russian government asked London and Paris outright to admit the priority of Russia's claims in an

*Memoirs of M. Sazonov, quoted by M. N. Pokrovski, President of the Academy of the USSR, in *Pages d'histoire, Editions sociales internationales,* Paris, 1929.

eventual dismemberment of the Ottoman Empire, with which, it should be borne in mind, nobody was yet at war.

The rulers of Russia were in that moment in high fettle. Petrograd and Moscow were covered with flags. Russia's armies had crossed the Oder, and the hard-driving Cossack divisions were reported "within five *étapes* of Berlin." On the other hand, the French and British were still falling back before the German onslaught in the west, and Von Kluck's Uhlans had photographed the outline of the Eiffel Tower from the heights along the Marne.

Now was the time, the Czar's cabinet decided, to make a frank avowal of Russia's war aims and force the Franco-British allies to accept the Czar's views on the future of Turkey. The demand for an immediate examination of the Eastern question was transmitted to London in a stiff note (the documents in the Russian state archives of the time say "ultimatum") which contained the threat of a Russian withdrawal from the war with Germany even before the Allies could carry out such a move, unless, of course, the Czar obtained satisfaction on the subject of Turkey and freedom of action to proceed aggressively against that country. London replied coolly on September 14 that Russia's interests in Asia Minor would be taken into serious consideration by His Majesty's Government . . . after the war.

This evasive reply hardly satisfied Nicholas II. And one cannot imagine how the Entente would have dealt with the problem of a Turkey, which was not a belligerent and which had clearly no desire to fight and, least of all, to be dismembered. The Germans, however, cut the knot for the Entente. The German diplomatic mission to Constantinople revealed the depth of the Russian intrigue to the already skeptical Turks and pointed at the same time to the perilous military position of the Allies. Not only were the Entente armies in the west in such a dire predicament that Britain was sounding President Wilson on the possibility of starting negotiations for an armistice, but Russia's hour of glory was also drawing to a close. The Turkish general staff was shown a plan of the preparations drawn up by General Ludendorff for a campaign in the East. According to that plan, the Germans intended to strike a blow that would knock Russia out of the war before Christmas and break her military power for years to come. With that evidence before them, the Turks were invited to join the side which was obviously going to be victorious. And the Turks accepted.

On October 29 they sent a naval expedition against the Crimea, bombarded the docks and arsenals of Novorossisk, and in one fell swoop crippled the Russian Black Sea fleet, driving its remnants into the harbors of Sevastopol and Odessa for shelter.

At last the Czar had his war with Turkey for which Russian diplomacy had never stopped plotting. Still he was not satisfied. In the Winter Palace and across the Alexander Garden, in the palace of the Holy Synod of the Greek Catholic Synod, there might be rejoicing over the fact that the Oriental question was about to be solved. But the British Government did not seem to share the satisfaction over the arrival of Turkey in the enemy's camp.

Contrary to expectation in Petrograd, the British did not make the Turkish declaration of war the occasion to open conversations on the military steps to be taken for the speedy defeat and dismemberment of the Ottoman Empire. Nicholas and his ministers, pursuing their fixed idea of loot in Turkey, did not take into consideration the serious position in which the new declaration of war placed the British Empire. For not only was the war on the Western Front going very badly, but in South Africa the Boers, under the old *voortrekkers* Christiaan de Wet and Jo Beyers, had risen in rebellion. On that account the Australian and New Zealand Expeditionary Force, destined to reinforce the hard-pressed British army in France, would have to be diverted to Capetown. Worse luck: the Turks, simultaneously with their naval raid in the Black Sea, had thrown all their Palestinian forces into the Sinaian Peninsula. In a week they had advanced to the shores of the Suez Canal and were reinforcing the Red Sea garrisons at Akaba, Wej, Yenbo', and Jidda. An able German strategist, Liman von Sanders, had been placed in charge of Turkish operations in the Near East, and the prospect of an invasion of Egypt suddenly stared the British in the face. They might be celebrating over in Petrograd; London took a more gloomy view of the successes of Russian diplomacy.

The Czar called in the French ambassador and spoke of his plans to make Germany—and Turkey—disappear from the map. The French diplomat was expected to relay this information to Paris and London and to assure the two Allied governments of Russia's determination to fight on to the bitter end. At the same time, however, Nicholas recalled Count Sergei Witte, a former foreign minister. Witte reappeared in the salons of Petrograd and with "a haughty and tranquil audacity," which threw the British and French ambassadors into consternation, preached openly the necessity of "liquidating the insane adventure," as he called the Great War and, in particular, Russia's participation in it. "Great anxiety" was the result when the British and French Government learned of Witte's declarations. For had not the man just been restored to favor by Nicholas II? And was it not the same Count Witte, "the navigator in the storm," who on a previous occasion had been recalled from disgrace to bring the war with Japan to a speedy conclusion? The British and French ambassadors were so alarmed that they asked for a private audience with the Czar himself.

In the meantime a British naval force was scouring the Aegean to watch for possible Turkish warships and transports carrying troops and equipment for disembarkation at the ports of Syria and ultimately destined for the army in Sinai. Reinforced day by day by units that became available as Britain cleared the oceans of German raiders, this naval force was to make an attempt to force the Dardanelles and bring Constantinople under the menace of its guns. Winston Churchill, then First Lord of the Admiralty, expected the Turks to sue for peace the moment they saw the Union Jack appear in the Bosphorus.

It was now the turn of Russia's rulers to be thrown into consternation. It seemed to them that Britain was about to steal a march on the Czar by attempting to force the Straits and seize Constantinople without

Russia's participation. Nicholas, whom most historians of the epoch present as a weak and vacillating person, spoke in unusually firm language when he received Maurice Paléologue, the French ambassador.

"To me," said the Czar, "the war against Germany and Austria and the alliance with France and Britain are but different means tending towards one single goal: I want Constantinople . . . . I do not hesitate to say however, that under the present circumstances Constantinople might well become the grave of our alliance if the Allies should do anything about occupying that city or regions adjacent to it without our participation. It is we who must have sole and absolute control of the Straits and of the city, and nobody in the world is going to interfere. My ambassador to Bulgaria, Prince Troubetzkoy, has sent me information which does not make me happy about the arrival of British naval forces in the Aegean . . . . The whole question interests Russian public opinion more passionately every day . . . . I cannot continue to impose the frightful sacrifices of the present war on my people if I cannot guarantee them at the same time as compensation the realization of their age-old dream. And so my decision is made: Constantinople and southern Thrace must be brought within the sphere of my empire. I am willing that some political regime be instituted in the city that takes into account the interests of other powers, but Constantinople itself is mine. That must be understood once and for all. Of course, I want France also to come out of this war greater and stronger. I agree in advance to all the French government's desiderata. Let France take Alsace, take the left bank of the Rhine, take Mainz and Coblenz, take anything you want, *monsieur l'ambassadeur,* but I must have Constantinople . . . . "

On March 10, 1915, the British ambassador to Petrograd, Sir George Buchanan, handed Sazonov a memorandum wherein Constantinople is called "the richest prize of the war" and wherein the Russian Government is informed that the Turkish capital will be hers after the war.

The British Foreign Secretary, Sir Edward Grey, who wrote the memorandum, made it clear that, in the view of His Majesty's government, the question of Constantinople involved the future of the whole Near East: Palestine, the Arabian countries, and even Persia. He insisted, although it was agreed by His Majesty's government to let Russia have Constantinople at the end of the war, that the regime to succeed the Turks' in Arabia be the subject of future discussions between the three allies: Russia, Britain, and France.

In return for the approval of Russia's designs on Constantinople and the Straits, the British government asked Russia to consent to the inclusion of the whole of central Persia into the British zone of influence.

Hearing this, Sazonov, without a moment's hesitation, said to Buchanan: "I agree!" And with these two words was solved the so-called "Persian question" which had kept Russia and Britain at loggerheads for two centuries.

Even so, the whole diplomatic transaction concerned what proved to be an academic question. The British were as far from Constantinople as were the Russians, and at the end of the war the Sultan still sat in his harem at Yildiz Kiosk. The British navy completely failed to force the Dardanelles on March 18, 1915. It was then evident that nothing but a siege on the grand scale would ever bring Turkey to her knees. Happily, Generals Botha and Smuts had crushed the Boer rebellion so that the Australian and the New Zealand army, instead of going to South Africa or France, was set ashore in Egypt and later, under Sir Ian Hamilton, landed on the tip of the Gallipoli Peninsula, with the object of marching north to Constantinople. This attempt, too, failed. The British army was driven back with great loss and re-embarked for Alexandria. By that time Hindenburg had annihilated Russia's first-line armies in the battle of the Masurian Lakes. The Turks had invaded the Russian Caucasus, and Grand Duke Nicholas, the commander in chief, was imploring Britain to do something, anything at all, to divert the Turks and keep them off his neck.

Thus, it may be said that Great Britain was forced or dragged into the war against Turkey by her imperial Russian partner. It may be argued, of course, as has been done on more than one occasion, that if the Russian Autocrat, as he said, saw in the Turkish declaration of war the pointing of God's finger to the dawn of a new era, the British cabinet, too, received the news of the occupation of Sinai by a Turkish army as the answer to its secret prayers. But this view cannot reasonably be maintained. For while Britain's interests might sooner or later have clashed with those of the Turks in that extremely inflammable neighborhood of the Suez Canal and the British Government could ill afford in the long run to leave that chief artery of empire under the physical domination of the Turks in Arabia, it is also true that the British genius for compromise and steering a midcourse might just as well have postponed an armed conflict indefinitely.

In 1914 the problem was not a pressing one. The Turks were by no means eager to bring the issue to a head. The members of the British cabinet could not in 1914 foresee the favorable outcome of a war with Turkey that subsequent events brought about. It is, above all, not reasonable to suppose that Britain seized with alacrity upon an opportunity to settle old scores with the Ottoman Turk at the precise moment when her armies in France and Flanders were fighting with their backs against Dunkirk, when her chief ally had begun to bleed to death on the battlefields of Verdun, and when the submarine campaign to force England to its knees was just getting under way. Britain accepted the war with Turkey as an unavoidable and terrible burden and did the best she could, once Turkey was in the war, to knock her out.

The failure to carry the Straits in March, 1915, must be attributed to Lord Kitchener's reluctance to make the action more than a naval demonstration and to his refusal to have the fleet's attack synchronize with a land operation. Gallipoli failed for the opposite reason: Sir Ian Hamilton's army of Anzacs and Jews was inadequately supported by the navy because the Admiralty feared to risk the loss of more ships and because the Turco-German allies had in the meantime poured large forces into the Gallipoli Peninsula and had reinforced both the European and Asiatic shores of the Straits.

In order to strike that blow which would ultimately

shake the Ottoman Empire to pieces, the British first had to withdraw, gather their full strength, and, like a jumper before vaulting the hurdle, measure their distance that still separated them from victory. The task called for long-range preparations and a start from scratch.

After throwing the Turkish army out of Sinai into southern Palestine and holding it there, the British High Command chose Mesopotamia for a jumping-off place to attack the Turkish Empire's eastern flank. This was another miscalculation that led to another false start. Originally, the expedition up the valley of the Euphrates was undertaken to throw a screen in front of the oil fields at the head of the Persian Gulf. It would have been an easy job for a small Turkish force to seize the installations, destroy them, and deprive the British navy of an important fueling station. The Turks apparently did not think of it. They seemed to have favored leaving well enough alone. Another consideration for marching up the valley was the fact that Britain had seventy million Moslem subjects in India who looked upon the Sultan (who was still Khalif ul-Islam) as the Shadow of God. The British army in Mesopotamia had to prevent the Turks from establishing contact via Persia with the millions of Mohammed V's spiritual followers in India. Pan-Islamic propaganda was said to have made enormous headway since the advent of the Young Turks. But the Turks did not think of rousing India against the British either. Nor did the Indian Moslems give the slightest indication that they objected to the British war on the Sultan, who was like a pope to them.

Pan-Islamism proved to be a myth, a wishful trumpery of Turkish clericalism, which was fighting for its bread and butter. Upon the insistence of the Germans, Mohammed V, as Khalif ul-Islam, did indeed proclaim a holy war against Britain. But there was no response whatever to his religious appeals. Theocracy was dead. The Sultan's call was no longer the call of faith. The old saying: *la wataniyah fil Islam*—there is no nationality in Islam—was no longer true in the sense that a supranational, mystic solidarity took precedence over local allegiances amongst the Moslem peoples. Imperialism carries the seeds of its own destruction in its bosom. It cannot but breed a fierce national chauvinism in the peoples it subjugates and exploits. Ottoman imperialism made no exception to this rule. Instead of rallying around the central power, some of the nationalities in the Ottaman Empire began to stir uneasily when Turkey became engaged in a major war.

The British army in Mesopotamia, composed of Indian divisions, at first advanced rapidly. It won battle after battle. Resistance fairly evaporated at the sight of the khaki-clad Sikhs and Gurkhas. General Maude, the commander, began to think that an Indian army was better than Turkish troops. Over in Britain, Colonel Lawrence says, the War Office was bursting with confidence. General Maude would not merely protect the Persian oil fields, he was going to march up to Baghdad, strike west from there for Damascus, and make for the Mediterranean shores. So they talked in Whitehall. Some local Arab tribes who offered aid to the advancing British were told to stay in their tents: Britain could do the job alone. There followed a rapid advance as far as Ctesiphon. Then General Maude was abruptly checked. He had met the first Turkish divisions, soldiers whose hearts were in the game. Then only did it dawn on him that those first enemy divisions he had met and who had melted before his advance like snow in springtime were entirely made up of Arabs. The Arabs had preferred surrender or flight to battling for their immemorial oppressors. At Kut the British expeditionary force was trapped by a real Turkish army and, after an agonizing siege, forced to surrender. Efforts to relieve the city cost twenty-four thousand casualties. Of the captured garrison of nine thousand few saw England again.

In fact, the circumstances of the fall of Kut, after a futile attempt to buy off the besieging Turkish commander with two million pounds sterling, form one of the most ghastly pages in the literature of war. A German officer, who was present at the surrender and who watched the British prisoners start out on the road to Baghdad, observed: "Not so many as ten out of every hundred will ever see their homes again." The men were given so little food that they sold their boots and clothing to the Arabs for a handful of dates and black bread. Soon they were incapable of marching in the burning heat. The march itself, says *The Official History,* was a nightmare. The Arab soldiery, who escorted the British prisoners, freely used sticks and whips to flog the stragglers on.

Major C. H. Barber, who accompanied the march, wrote:* "We tingled with anger and shame at seeing a

---

*Quoted in *Loyalties,* by Sir Arnold Wilson, Oxford University Press, London, 1930.

column of British troops . . . driven by a wild crowd of Kurdish horsemen who brandished sticks and whips. The eyes of our men stared from white faces, drawn long with suffering of a too-tardy death. As they dragged one foot after another some fell, and those with the rearguard came in for blows from cudgels and sticks. I saw one Kurd strike a British soldier who was limping along, he reeled under the blows . . . . Another was killed outright with a stirrup iron for stopping a few seconds on the road . . . . Their feet were a mass of blood . . . . Some were thrashed to death, some killed, and some robbed of their kit and left to be tortured by the Arabs. Men were dying of cholera and dysentery and often fell out from sheer weakness. Every now and then we stopped to bury our dead . . . . Enteritis, a form of cholera, attacked the whole garrison after Kut fell . . . . A man turned green and foamed at the mouth. His eyes became sightless, and the most terrible moans conceivable came from his inner being . . . . They died, one and all, with terrible suddenness. We saw officers and men lying uncovered from the sun on stretchers covered by thousands of flies. Now and then a wasted arm rose a few inches as if to brush them off but fell back inadequate to the task. Emaciated corpses, stripped of their clothing by the Arabs, lay around every little pool of water . . . . One saw British soldiers dying, with a green ooze issuing from their lips, their mouths fixed open, in and out of which flies walked . . . . Thousands of the British prisoners were never heard of again . . . . Officers were deliberately starved to death . . . . Groups of men were left for the Arabs to play with, torture, and mutilate . . . . "

Notwithstanding its failure, the Mesopotamian campaign had its compensations. It showed the Duke of Wellington to have been right when he told the British cabinet that a great nation cannot fight a little war. The dash up the valleys of the Tigris and Euphrates brought home two things; first: that the Sick Man of Europe was by no means as decrepit as diplomatic and other rumors had made him out to be, and secondly: that there might be a possibility of stirring certain Arab tribes to revolt against their Turkish overlord. The British government in India, which had and still has its spies amongst the pilgrims who annually visit Mecca, pointed to Abd al-Aziz Ibn Sa'ud of the Nejd as the strongest of the Arab princes and as the most likely leader of a revolt against Turkey because of his puritanical principles, which clashed directly with the religious laxity in the Sultan's household and in ruling Turkish circles.

The British invited Ibn Sa'ud up to Basra, where he was amply made to feel the importance they attached to his eventual collaboration. For the first time the desert chief played the role of a big potentate, receiving salutes, inspecting naval sloops, guards of honor, and docks. He acquitted himself of his duties with the easy nonchalance of a practiced monarch. Gertrude Bell, H. St. John Philby, Sir Percy Cox, and other British observers and agents lauded him as a lion in battle, a lamb in society, and an angel in council. His military fame and statesmanship were compared with that of the khalifs of Islam's golden epoch.

At first the Wahabite chief seemed inclined to accept the leadership of an eventual Arab revolt that the

British suggested. He knew enough about the railway schemes of the Germans to realize that once the Baghdad line reached the Persian Gulf his days as an autonomous chief were numbered. His fundamental idea was, and remains till this day, to keep the Arab world untouched by modern progress and unsullied from contact with the Christian infidels. In his own realms he has strictly adhered to this policy, accepting only what technical Western aid was necessary for the exploration of oil fields, but for the rest shunning contact with what he called once "the poisonous Occident."

He naturally fell in, therefore, with the British plan to bar further German penetration of the Arabic Peninsula and he demanded that Britain, if victorious, would pledge herself not to take Germany's plans of covering the Near East with a net of railways of her own and otherwise push the modernization of the Arab world too fast. This promise was made readily enough since Britain herself definitely preferred the status quo of semi-feudalism to progressive innovations.

However, Ibn Sa'ud had a rival, Ibn Rashid. Him he desired to eliminate before initiating the general movement of revolt. He mobilized his army of tribesmen and attacked at once, but received so severe a drubbing that he recoiled from undertaking any major struggle thereafter. Captain Shakespear, the British officer, who had carried on the negotiations with Ibn Sa'ud and who served as the Emir's military adviser, was killed in the battle with Ibn Rashid. But for this tragic accident, to quote St. John Philby, Colonel Lawrence might never have had the opportunity of initiating and carrying through the brilliant campaigns with which his name is associated. The world would have been deprived of an epic, the British treasury would have saved hundreds of millions of pounds sterling, and the family of Sherif Hussein of Mecca might never have emerged from obscurity and from the arms of the Turk with whom they were in active negotiation until April, 1916.

After Captain Shakespear's death, Ibn Sa'ud refused to stir. He suddenly developed religious scruples. To him the Sultan, in spite of all his faults and his impious environment, was still the Commander of the Faithful. Hearing the imams cry out three times a day the Prophet's words against the unbelievers, Ibn Sa'ud could not bring himself to enter into an alliance with them, however secret it might be for the moment. The British had to look elsewhere for support in the great project for ousting the Turk from Arabia.

There were hopeful rumors, on the other hand, of certain Arab divisions in the Turkish army said to be seething with discontent. But they were nearly all stationed in the vicinity of Damascus. Other divisions said to be similarly disaffected were camped in the neighborhood of Baghdad. There were also the secret revolutionary committees in all the great cities of Syria, and they had sent agents to the British in Egypt to tell them of their willingness and readiness to raise the green banner of revolt.

But how could Britain establish contact with all these potential allies and supply them with arms and equipment? They must not be allowed to fling themselves into an insurrection without due preparation and before the British could appear on the scene in force. That would be wasting mighty potential energies which, under different circumstances of organization and co-ordination, might later bring success. The Turks were certain to drown in blood any incipient revolt. The Turkish military governors, who ruled in Baghdad and Damascus, were iron disciplinarians. They had learned the lesson of massacring civilians in Armenia and Kurdistan.

While negotiations and discussions about an Arab revolt were going forward in London and Cairo, Sherif Hussein of Mecca, the custodian of the holy relics and an authentic descendant of the Prophet, who, like Ibn Sa'ud, enjoyed a position of semi-independence under the Turks, offered his services. He had been in communication with the British before the war broke out. From him, indeed, had come the grave intelligence about German preparations to install submarine bases on the Arabian shores of the Red Sea.

In July, 1915, the Sherif of Mecca solicited His Majesty's government's help for the cause of Arab independence. He asked for himself, to begin with, a donation of one hundred thousand pounds sterling and, after the successful termination of the revolt, sovereignty for his house over the whole Arab-speaking world in southwestern Asia and northern Africa. He claimed he had influence everywhere and everywhere the men ready to do his bidding. At the British Agency in Cairo, Sir Ronald Storrs, to whom fell the decoding and translation of the Sherifian message, could not help muttering the old refrain: "In matters of commerce the fault of the Dutch/ is offering too little and asking too much," as he read the Sherif's preposterous pretensions.

Hussein's claims were indeed fantastic. He did not seem to take into account that the British knew perfectly well that if he started a revolt against the Sultan, his liege lord, he would be considered a traitor to the Shadow of God by at least ninety per cent of the Moslems throughout the world. His immediate neighbor, the Emir of Nejd, Ibn Sa'ud, wasn't his friend and ally at all, as he claimed, but a deadly foe who in time was to dethrone him and chase him into exile. Another neighbor, Yahya ibn Muhammad, Imam of Yemen, was known to speak of the Sherif disrespectfully as "a piece of vileness left in Mecca by the Turks."

Then there were the northern populations of Palestine, Lebanon, and Syria, who certainly did not look for leadership to an official of the Sultan in a backward bailiwick like the Hejaz. To the great Arab peoples of Africa (most of them Shiites), whom Hussein claimed as his devoted followers, he was simply a Sunni heretic.* But he had one great real advantage over all other Arab princes. He had a son named Feisal. Him the British knew to be a man of genius and a born leader. Feisal was educated in Constantinople and had traveled widely in Europe. His fame had spread through the whole Arabian world. It was said of him that many a time did men pray that they might behold his countenance. When the British learned that Feisal would join his father in a revolt against the Turk, they accepted

*Sunni, Shiah, and also Khawarjii, are the most important sectarian divisions in Islam. They have their origin in divergent theories on the office of the Khalif, the head of the Muslim community, as successor of the Prophet.

the old man's offer in principle. But they did not commit themselves to yield to his extravagant demands and were kept fully informed of the secret negotiations he was to carry on for another year with the Turks.

For all that, the Arab revolt never came off. In spite of Feisal's unquestioned sincerity and political vision, in spite of the tons of gold that Britain poured into Hussein's seemingly bottomless exchequer and into the outstretched palms of a hundred Arab chiefs and chieftains, in spite, too, of the account of T. E. Lawrence's exploits and experiences given in *The Seven Pillars of Wisdom,* and in spite of everything else that has been said and written on the subject and that is still being said (for the myth of the Revolt in the Desert is part of the basis on which certain Arab politicians hope to build a new political structure in the Near and Middle East), the Arab rising never took place. In the annals of history the Revolt in the Desert must be placed in the same category as the Trojan War. Lawrence is the Homer of that latterday Semitic Iliad. It is legend pure and simple, a convenient legend no doubt, as many legends are, and in spots most beautiful and inspiring, but it no more deserves a place in the frame of historical reality than the story of the Angels of Mons, whose swords flaming in the night are said to have saved the famous Contemptibles from annihilation.

At Mons the British soldiers themselves fought their way out of Von Kluck's deadly grip. In the Near East it was the blows delivered by Allenby at Gaza and Megiddo which loosened the hold and broke the might of the imperial Ottoman and sent him reeling back towards Damascus, Aleppo, Homs, and Antioch into the trackless wilderness of Adana, where no man could pursue and where it was not necessary to pursue. For the capture of Damascus marked the end of the war. After Allenby's occupation of that city Turkey capitulated, soon to be followed by her Austro-German allies.

In those operations of Herculean magnitude—the assembly of the Army of Egypt, the laying of the water pipes and the building of the supply system, the thrusting back of the Turks from the Suez Canal, the march from Sinai into Palestine, and the capture of Jerusalem, Jericho, Haifa, and the Plain of Jisreel—in all these activities the irregular force of Arab tribesmen which had been drummed together by Feisal and Colonel Lawrence played no role whatsoever. It roamed up and down in the open spaces of the desert to the East, damaging the Hejaz Railway here and there, but never seriously enough to interrupt traffic for more than a few days. After two years of shouting, riding, and quarreling, in the course of which the throats of some isolated Turkish patrols were cut, the tribesmen did not have to their credit the capture of a single place that the Turks had decided to defend and hold. The city of Medina, which, so to speak lies within a stone's throw of Mecca and Jidda, the starting point of the so-called Revolt, held out till the spring of 1919, long after the Armistice had been signed, and only then surrendered on specific instructions from the Turkish government to the commander of its garrison.

Lawrence relates with dismay and disillusion how every relatively important operation entrusted to the

Arab irregulars, every task of some potential consequence set before them by the British High Command, miscarried or went awry. In his great book, the like of which is written but once in a generation he recalls the sense of despair and futility that assailed his spirit and came near to overwhelming him more than once in the long months of exhausting effort to whip the Arabs into some semblance of a cohesive fighting force. When a Turkish train was wrecked by dynamiting some culvert, bridge, or trestle, the tribesmen pranced around in childish glee at the sound of the explosions and then rushed forward, if the train crew and convoying military were too stunned to resist, to gorge themselves on whatever food they could lay their hands on and then streaked back to their villages, their camels loaded to the breaking point with loot.

Seldom did the Turks give pursuit. Had they done so, or rather had the Turkish High Command thought it worth the trouble to send an expeditionary force south along the Hejaz Railway, there is little doubt that short shrift would have been made of Feisal's army.

At no time in the course of Allenby's campaign did the Arabs immobilize more than two or at the most three of the skeleton Turkish divisions dispersed over the blockhouses and posts guarding the railway from Damascus to Mecca. Feisal's procession through the desert looked as picturesque and terrible as an army with banners, but the banners wrought no destruction. He called himself "a preacher of the good news." His mission was preaching, not fighting. Feisal never came to grips with the enemy in a moderately large-sized pitched battle. He raided, harassed a Turkish post, or damaged a section of the railway, but if the Turks came up in force, he hastily withdrew. Every engagement the irregulars undertook was planned in the most minute detail beforehand to make it a complete surprise to the Turks. For it was understood—Lawrence makes this quite plain—that there were to be no casualties on the Arab side. A single engagement with heavy casualties, he writes, would have sent the remnant of Feisal's army home, and no amount of reasoning or inducements could have kept the tribesmen under arms an hour longer.

Some of the strongest Arab tribes refused to have anything to do with Feisal's grandiose projects of which Lawrence was the theoretician and apostle. They looked upon the movement of liberation as an English scheme to transfer sovereignty over their lands from Turkey to Britain. Deadly intertribal feuds kept rival clans at home lest a single day's absence of the fighting men bring their hereditary foes into their encampments to loot and wreck. Other tribes were with Lawrence and Feisal one day and with the Turks the next, depending on whose guns were nearest to their settlements or encampments. The combination of tribes into one great striking force was impossible because of their mutual distrust and hostility. The principal Arab chiefs, the rulers of Nejd, Asir, Yemen, and Hadhramaut did not deign to take notice of the movement, although there wasn't a Turkish soldier in their domains to prevent them from joining "the national cause." The Iraqi Arabs did not stir till the end of the war, after the capture of Baghdad by a British army and the Turkish surrender.

Palestine, some of whose Arab inhabitants are now loudest in claiming fantastic achievements in the war of liberation, remained within the Turkish lines till the end and never gave their overlords the slightest trouble. The Turks were not expelled by General Allenby until a few weeks before his final dash for Damascus in the late summer of 1918. Furthermore, Palestine before the war counted militarily even less than economically. It was a wilderness strewn with ruins. Its inhabitants, half a million or so poverty-stricken peasants and Bedouin, were of all the Arabs held in lowest esteem by the Turks. Only the Jews of Palestine were on Britain's side, and the Jewish Legion, recruited in Western countries, fought in the British army.

Not only did the Palestinian Arabs not rise in revolt but they assisted the Turks to the extent of waylaying and murdering British patrols and individual soldiers who strayed off the main line of march. Their leaders, all members of the large landowning families, the Mufti of Jerusalem Haj Amin Husseini included, served as officers in the Turkish army, a circumstance that has never prevented these effendis from claiming a share in the war of independence and, of course, compensations and concessions commensurate with such generous patriotic conduct.

Not till Allenby advanced from Gaza to Jerusalem and Jericho and then suddenly in one bound leaped for his final objective—Damascus—did the Arab world come in motion. Then the influx of tribesmen willing to share in the loot and plunder became so great in Feisal's army that food supplies threatened to run out.

But then it was August, 1918: the war was over, and Colonel Lawrence rushed off to Egypt in disgust . . . .

In Syria alone, of all the former Turkish domains, does there seem to have been a genuine revolutionary movement which dated from 1908, the year in which the Ottoman Empire, after six hundred years of glory and conquest, began to shake with the rumblings of its impending doom. But the Syrian revolutionary committees rendered little service to Allenby and the British army beyond being on hand to welcome them upon their entry into Damascus.

The myth of the Arab Revolt did not come into existence until Lawrence published in 1926 *The Revolt in the Desert,* which was an extract from *The Seven Pillars of Wisdom* that followed later. It then took shape in the minds of certain Arab nationalists and their well-wishers as a convenient ideological weapon against the settlement of Palestine by the Jews. Lawrence did not create the myth of the Revolt. His book proves that there was little more than a murmur in the desert, punctuated by a few explosions. On the last page he claims to have set a wave of Arab nationalism in motion and that this wave broke and washed away at Damascus. It is true that he raised some hopes. But he did not have to fire a shot to enter the city. The Turks had left, fearing the arrival of Allenby's army. Allenby's campaign ushered in the first faint glimmers of Arab nationalism. Of this campaign the Revolt in the Desert was an insignificant, ineffective sideshow; it did not make the slightest difference in the final accounting . . . .

In 1927, when I joined the correspondents of the

Berliner Tageblatt and the *Vossische Zeitung,* Dr. Wolfgang von Weisl and Ernst Davies, in a tour of the Arabian Peninsula, we soon convinced ourselves, without suspecting that several military historians, both English and foreign, had reached this conclusion before us, that the Revolt in the Desert may indeed have been the most hilarious, whacking, yelling, bubbling, colorful gun-firing Fourth of July jamboree that ever moved through the wilderness, but that its military value had been well-nigh zero. It was an idea, a stunt. Feisal pretended that he would revive the glories of the Umayyads and Ayyubids, but the Arabs remained skeptical to the end. He did succeed in uniting a few of the inland tribes for a moment by instilling in them an enthusiasm for Damascus and loot and women, but with Allenby's victory the movement burst like a soap bubble. The tribesmen showed no more aptitude or desire for ordered government than they had showed stomach to fight for it. In their advance through the empty spaces they moved without haste. At every week's long halt they made new claims on Feisal, dividing the bearskin a hundred times before the animal was caught and then in the end allowing someone else to take it. The chiefs were suspicious of each other and quarreled as to who should have the greatest honor when success should finally come.

Had Britain not reinforced the cavalcade with some detachments of Indian gunners and Sepoy infantry, the insignificant local successes would have been lesser still, if not non existent. A week or so before the goal was in sight, before Allenby moved in force upon Damascus, Lawrence could still think despairingly that the Revolt would never get through its last stage, but would "remain one more example of the caravans which started out ardently for a cloud-goal, and died man by man in the wilderness without the tarnish of achievement."

The raid on Atwi which Lawrence mentions was not a battle. We found two stone buildings there that once housed the Turkish garrison. The steel shutters still bore the marks of bullets. No more than fifty men, if they were packed as tightly as sardines, could have found shelter in these structures. They were overwhelmed by a mob of three thousand howling dervishes and massacred. Such was a typical victory in the Revolt. Of course, it meant something to the Turks; the loss of telegraph wire, tools, spare rails, and men. But Turkey held everything south of Deraa very loosely anyway. Jemal Pasha tried to keep the line open because of the garrison in Medina, which never surrendered, no matter how many times Lawrence blew gaps in the railway.

The tribesmen dynamited some wells and cleared out others which the Turks had tried to demolish. They surprised a few desert patrols. We saw Ghadir el Haj, the first station south of Maan and scene of a battle in the Revolt, and talked with the Turkish stationmaster. He remembered the battle very well. Yes, Feisal's men had made great ado there: they slaughtered two thousand sick baggage camels the Turks had pastured in the plain after the first battles in Palestine.

And so everywhere: they had thrown panic into the blockhouse posts, but had seldom attacked them. Mriegha and Waheida, which we inspected, were garrisoned with thirty men apiece at the time of the

Revolt. They were taken, only to be reoccupied by the Turks a week later. Feisal and Lawrence had no strategic plan, no guns, no communications, no money. They roamed hither and yon and were immobilized completely during the long winter months of the only year they were in the field. The Turks did not budge till Allenby's army to the west burst through the Valley of Esdraelon into Galilee and the Hauran. Then they knew the game was up. But the Revolt in the Desert had nothing to do with that.

We were satisfied that the Revolt in the Desert is literature, great literature, one of the finest books ever written, and then left that spurious trail.

We journeyed, by military motorcar, from Jerusalem to Beersheba, then into Transjordania, touching at Maan on the Hejaz Railway, back into Palestine, and on to King Solomon's ancient port on the Red Sea, stifling and dilapidated Akaba, in whose harbor once lay the fleets of Ophir and Ethiopia carrying gold, silver, horses and timber, inconsiderable now but a city that will regain its importance when, in the process of the East's rejuvenation, the modern Holy Land's industrial development orientates itself, as it inevitably will, towards the immense, British-held virgin markets of East Africa. Starting from Akaba, where Albert Londres of the *Petit Parisien* joined our small expedition, we first explored the Sinai Peninsula, visiting the Franciscan hospices and monasteries on the burning-hot mountains and in the furnace-like, flint-strewn valleys where Moses received the divine revelation and the Israelite nomads wandered before filtering into their

Promised Land. We duly sampled the alkaline waters of Meribah, saw patches of the mysterious manna on the ground, glistening like coriander seed under the shrubbery at dawn, went through the deserted, rock-hewn city of Petra, and in the stillness of the night imagined that most awful procession of history, with the Ark of the Covenant in the lead and a column of fire preceding it, bearing down on us out of the moonlit mountain haze.

Then we struck out for Jidda to visit H. St. John Philby, the adviser of King Ibn Sa'ud, who had sent word that he could not guarantee our safety in Mecca but would welcome us to Jidda instead. Twice, once to go and once to return, we followed the blinding sand-trail along the eastern shore of the Red Sea, where the sun cuts like a two-edged sword, to the city where Allah's vicar, Abd al-Aziz ibn Sa'ud, permits foreign consuls to dwell. There we learned, officially, of the Wahabite king's sympathy for the Jewish settlement in Palestine and of his fierce hatred for its chief opponent, the Mufti of Jerusalem, "that dog of an idolator and Satan's spawn, whom His Majesty would willingly chase into hell." Ibn Sa'ud's Wahabism, rigidly puritan, which bans smoking and barbershops and forbids women to draw man's attention in the streets, by the tinkle of their hidden ornaments, looks with reproach on all believers who worship in mosques. His is the creed of the open spaces. That is why he ruled out the Jerusalem Mufti. Reaching Maan again, we followed the old pilgrim road through Transjordania into the Hauran, leaving the evangelic Sea of Galilee on our left. Then we went through the Djebel Druse country to Damascus, Homs, Hama, Aleppo, and Alexandretta on the border of Turkey, returning along the coast to Beyrouth. Then we faced east to Mosul, south to Baghdad, past the ruins of "the Chaldees' excellency," following the course of the mist-covered Euphrates to Basra and so to Kuwait on the Persian Gulf, where we took ship for Aden.

From Aden, which is, I think, after Djibouti the hottest place on earth, we were permitted to visit primitive Sana, the capital of Yemen, returning to the Red Sea coast at Hodeida, to be picked up by a Greek tramp steamer which dropped us at Suez. From there we crossed Egypt's eastern desert in a final three-hour dash through infernal heat and settled in Cairo to write dispatches.

The notebooks containing my day-to-day observations on that weird fourteen weeks' ramble were left behind in France when the German enemy approached Bourg-en-Forêt, but the chief impression of Arabia gained in that year vividly remains. It is one of the immeasurable vastness of that huge quadrangle which is designated on the map under the general name of Arabia and of the perplexing heterogeneity of its inhabitants. If, upon entering that area, be it at its northern extremities, near Aleppo, for instance, or in the extreme South, let us say at Aden, the traveler still entertains some notion that he is about to traverse one country—Arabia—and meet there one people—Arabians—he will be undeceived before he has proceeded half a hundred miles inland. The distances between the first ten cities he encounters and the sober variety of the scenery and of the natural phenomena, whether he goes from north to south or vice versa, will soon dispel from his mind any preconceived ideas of one country inhabited by men of one race.

In every day's advance the impression will gain upon him that he has entered upon nothing less than a subcontinent peopled by races, nations, religious communities, clans, tribes, jarring sects and classes as different and widely apart as to mores, customs, appearances, and general cultural status as the nationalities of Europe. Arabia is a world in itself, as Europe is, or India. One would almost say, if it were not for a common religious bond which provides a certain similarity in dietary and ritual observances all over the Arabian Peninsula, that there is as wide a difference between Homs in Syria and Sana in Yemen, on the one hand, and as there is between Oslo and Sofia, on the other, though the two European and the two "Arabian" cities are approximately equidistant.

Homs is a city of the Levant, cultured, soft, and Hellenist. Sana is an incredible wilderness of dustheaps steaming in a blazing sun, inhabited by ferociously austere desert men. I shall ever remember Sana as the city where we saw entire families—fathers, mothers, and children—blind, forcing their way through the multitudinous bazaar hand in hand, to the monotonous, pitiful cry of *a'awar, miskeen*: I am blind, have mercy. Beyrouth is Constantinople in miniature; brothels, dance halls, hotels, soldiers, and French literary societies. Jidda is a squalid town of narrow streets, buildings five and six stories high, with latticework windows in the upstairs apartments, whence a thousand female eyes follow the stranger as he gasps for breath and staggers with the giddy heat in the streets below. When there the sun went in hiding behind a sad billowy sky, the humidity turned to a sulphurous foulness unimaginable.

But there, at last, we also found Philby, the man who had explored the Empty Quarter, that land of hunger, blood, and desolation and who wrote a book about it that ranks with Charles Doughty's immortal *Arabia Deserta.* Back in Jerusalem and Cairo they had referred to him in mysterious tones as the British agent in Arabia. Here he was held in universally awesome respect as the "Arabian Caesar's" adviser and personal friend. Having embraced Mohammedanism, this bearded Englishman had free access to the holy city of Mecca. But all our pleas to let us proceed thither if only for a day, and even if we grew beards, donned Arab garb, and learned the Koran by heart, fell on deaf ears. He ruled out Medina, too, and Taif and el-Lith and half a dozen other towns, peremptorily and with a gesture of uncontradictable finality.

This "arrogant British definiteness" so exasperated our friend Londres that he began to ask questions. Why should a private Englishman, Londres had wondered many times on the road before, have the final say over the movements of foreigners, of a French citizen at that, in an alien land? How illogical! How preposterous!

"Monsieur Philby, what really is your official standing or position here?"

"Oh," drawled the Briton, who had caught the undertone of querulousness in Londres' voice, "I merely represent a British commercial house here."

"What do you sell?"

"Sell? Sell, you mean? Oh, perambulators and sewing machines and that sort of thing, don't you know!"

"Monsieur Philby, when I have finished this jaunt," began Londres again, "I should like to visit Hadhramaut. To whom should I apply for a visa? They could not do anything for me in Egypt . . . . "

"For Hadhramaut, let me see now. For Hadhramaut, I think, you must make personal application to the British consul in Basra."

Londres gasped and knit his eyebrows.

"But Hadhramaut is an independent sultanate," he objected. "What has the British consul in Basra to do with it? Basra is an immense distance off. I can't go there now. Perhaps I had better visit the Pearl Islands . . . "

"For a permit to visit the Pearl Islands," came back Philby imperturbably, "you should apply to the British consul in Beyrouth . . . "

"In Beyrouth, the British consul in Beyrouth. But what *sacré nom d'une pipe,* has the British consul in French territory to do with the Pearl Islands? There is a king there, in the Pearl Islands. They are independent territory. They are . . . . " Londres grew excited.

"Oh, how like a Frenchman you talk!" exclaimed Philby.

Damascus, oldest city on earth, is a garden, as it was in the Prophet's day, still the most marvelous on earth, luxuriant, cool, and fertile, with running water aplenty and shady groves. At Kuwait one is surprised that such a collection of mud hovels and corrugated iron sheds bears any name at all. Nekhl, in Saudi Arabia, is no better. There are regions in Arabia, as in the neighborhood of the godforsaken ruins of Babylon and Nineveh, lizard-covered and infested with jackals and other diabolical creatures, where life is scarcely above the level of that of the most primitive tribes in Nigeria or along the upper Congo. In fact, the edge of the whole inland plain, itself fairly prosperous farm land, is peopled with tribes who live no one knows how or why, wasted by drought and disease, subject to murderous raids by Bedouin whom nobody controls and who among themselves carry on blood feuds from one generation to another.

The great cities are all in the north, but they are more Greek, Turkish, Armenian, and French in character than Arabian. They bear the mark of every civilization that passed over them: Assyro-Chaldea, Phoenicia, Hellas, Persia, Rome, and Byzance. They are cosmopolitan in feeling and therefore bereft of any aggressive idea of nationality. The triangle of which Alexandretta, Aleppo, and Homs are the cornerstones and in the middle of which lies the small town of Antioch, once the fertile source from which Christianity poured its vitality over Europe, culturally faces the West and is open to the West, even as Mesopotamia faces the East towards Persia and India. The Arab is a carrier of wood and a hewer of stone in northern Syria and Iraq. He is the peasant, the ruralite. The people call themselves Syrians and the intellectuals and businessmen speak in biting condescension of *"les Arabes là-bas,"* the Arabs down yonder, in the lower regions of the Peninsula. In these cities, as in Turkey, the Moslem religion is fading out to a mere set of moral rules, lacking all the dynamism and passion of former days.

East of the valley which contains these cities and olive groves of wheatland and patches of green sesame and alfalfa as well, all the way to Mosul and Baghdad along the Anatolian border, the language is Turcoman and Armenian. The peasants are Moslems with small settlements of Armenian Christians dangerously interspersed amongst them. But the coastal regions are peopled by Nosairis, who are pagans pure and simple, hating Christianity and distrusting Islam. They speak an Arabic dialect but write their idiom in Greek letters. Closer to the cities, under the immediate protection of the former Turkish governors and their guns, now departed, are communities of Circassians and Kurds, Moslems too, but still considered aliens and interlopers. Their language is Arabic, or rather an Arabic dialect, quite similar to that spoken in the neighborhood and surrounding environments. Their religion is Mohammedan like that of their neighbors, yet not a man among them will venture a step beyond the communal bounds for fear of assassination. It was in this neighborhood that we encountered a caravan of dead men, corpses bundled in white, strapped in the saddle of the groaning camels, on their way to burial in some particularly holy ground.

There is perhaps no other like-sized area in the globe which is torn by so many contrasting loyalties, deadly feuds, and irreconcilable animosities as the northern parts of Syria and Iraq. A man will guide you for a distance and be the most genial companion on earth until he comes to some landmark, unimportant to you or even unnoticed: a pile of stones, a dried-up creek, the bleached skeleton of a camel, a cluster of date palms, or possibly the faint outline of some huts visible upon the distant horizon, when he will suddenly announce that he cannot go another step. Entreaties, love, or money are unable to shake his decision. He solemnly informs you that he has reached the boundaries of a tribe with which his own clan has a blood-feud. You ask how long the mortal tension between the two has lasted and if perchance there be a possibility of a reconciliation through the friendly intervention of a stranger. But he replies with a hopeless sweeping gesture that what has been in all eternity will endure till the end of time.

We asked the Foreign Minister of Saudi Arabia about the prospects of an Arab federation and Arab unity and were told in a rasping voice crackling with harsh consonants and explosive aspirates that there can be no question of union or unity until the heretics have been exterminated or converted by the sword. We learned with amazement that by heretics were not meant the non-Moslems, the infidels of Europe or the Jews, who were considered beyond redemption anyway, but His Excellency's own kinfolk the Arabic inhabitants of Trans-Jordan, Palestine, Iraq, and Egypt.

Hearing this, Ernst Davis inquired boldly: "Then what are you waiting for to convert them to the true faith?"

"Dans la plénitude de temps, in the fullness of time," replied His Excellency, "there will arise a new Prophet, if Allah wills, who will bring all the peoples of the earth to the knowledge of truth. He may be born

now . . . . We do not know . . . . "

"By the sword, Your Excellency said? How will the many who are slain come to know the truth?"

"Allah will know his own!" came the sure reply, the selfsame answer the Duc de Guise made on the eve of St. Bartholomew.

In Jidda too, as in Damascus, we heard it said solemnly that the English would not forever treat Arabs as they treat the people of India, and our informants let us know that they were perfectly well aware that Feisal and Abdulla ruled over mock kingdoms, and that British preaching of federation was mockery supreme.

Damascus is the most intensely Arab in feeling. It is, next to Cairo, the most important center of Koranic theology and jurisprudence. It would be the natural capital of an Arabian Empire, if ever such a structure were reared, because since the Prophet's own days the eyes of all Arabs are turned in its direction for political inspiration. It is far more the center of Arab nationalism than Mecca, or Baghdad, which lost its leadership with the death of Harun al-Rashid and never regained it. Cairo never indulged in imperialist Pan-Arab speculations until very recently. Previously, in the days of Fuad I, the sheik-president of Al Azhar University and Egyptian statesmen and intellectuals indignantly denied to me in 1929 that they were Arabs at all. They said they were Egyptians first, last, and all the time. In Damascus a stranger is made to feel the pride of Arab nationality; in other towns and cities he is tolerated as a religious nonconformist and innerly despised. Arab affairs are the chief subject of conversation in Damascus. In Aleppo and Alexandretta it is French novels, French music, and French parliamentary vicissitudes and personalities that occupy the intelligentsia in their apparently endless leisure hours. Beyrouth is *une salade,* a melting pot of races which do not melt but which fidget and shout and argue and fight in bastard French. The vehemence of the language of the half-dozen French Journals is reminiscent of the Parisian *faubourg* at election time. The town swarms with secret societies, revolutionary clubs of Syrian, Greek, Lebanese, Russian, and Armenian variety.

The peasants outside Beyrouth in sweetly named Lebanon right down to the Palestinian border and the urban *petite bourgeoisie* are Maronites and Greek Orthodox in religion with a sprinkling of nonconformist groups such as Baptists, Seventh Day Adventists, Methodists, and even, to my immense delight, Dutch Reformed. They learned the denominational splitting-up in America, where many of these "Syrians" have lived and prospered for a time. They are united in one thing: their fear and detestation of the Moslems. Religion is a substitute for nationality all over the Peninsula. But their villages look neat and prosperous, and their farms are well kept. Apple orchards, rosemary, melon patches, date gardens, and olive groves make the Lebanese landscape a pleasant mosaic. Next to Jewish Palestine, Lebanon's agriculture is the most progressive in the Near East. There, too, we made a special ascent of the Anti-Lebanon to touch with our hands the last remaining cedars from whose wood Solomon's temple was built. Of these massive trunks, the contemporaries of the Song of Songs, few remain.

In the hills of eastern Syria, where both the Tigris

and Euphrates rivers originate, the predominant religion amongst the settled tribes is Shiite Mohammedanism. The peasants came generations ago from Persia, where that sect is supreme. In Syria they consider themselves a race apart. They are in exile, strangers in a strange land. Like Ishmael, their hand is against everyone, and everyone's hand is against them. But their chief hatred, they are not slow in telling travelers, is directed against the Sunnite Moslems, whom they detest more bitterly than Christians, which is saying a good deal in a part of the world where a Christian is held in lower esteem than a dog and is, in fact, frequently called by that name.

From Damascus southward to the Galilean borders of Palestine, the country is inhabited by the fierce and backward Druses, a warrior race who are heretics to all other Moslems. I saw a good deal of them in 1925 and 1926 during their revolt against the French mandatory regime. They are followers of one al-Hakim bi'amrillahi, the son of a Russian mother, who proclaimed himself an incarnation of God in the year 1016. This al-Hakim was an Egyptian who never visited Syria, but who sent his missionaries to that country of receptive moods. They converted the Druses to a belief in Al-Hakim's divinity. The Druses are not Arabs, but Kurds and Armenians by descent. The Druses are mountaineers for security's sake. One of their settlements lies on the western slope of Mount Hermon, the snow-capped watchman of Palestine. There are Druses in the Holy Land and several thousands in the United States of America, where they pass for "Syrian" Christians, the curious part of the Druse religion being that a follower of al-Hakim is permitted to conform outwardly to the faith of the unbelievers among whom he dwells. On two occasions in modern times the Druses brought themselves to the notice of the civilized world: once in 1860, when they went through Lebanon slaughtering tens of thousands of Christians and sacking the foreign consulates in Damascus; the second time in 1925, when they rose in revolt against the French. Against the Turks they rebelled a dozen times. Mixed with the Druse villages are settlements of a nondescript rabble, neither Druses nor Bedouin, who have an obscure cult of which John the Baptist is the center. As secretive about their religion as the Samaritans near Nablus and like them few in number, they are dying out because they refuse to be contaminated by intermarrying with lesser breeds.

In the Hauran district the mountains of which are occupied by the Druses, and in the Yarmuk Valley lives the sturdiest peasant race of all Arabia, prosperous, good-natured, and loyal. On the east bank of the Jordan, however, their villages run into the settlements of the Algerians, who are universally hated as intruders, in spite of their century-long residence. Other tribesmen warn the stranger that the Algerians do not even recognize the sanctity of the visitor's life in their houses or tents. The Algerians are descendants of the irreconcilables from that north-African country which was lately occupied by General Eisenhower. When Algeria was conquered for France in the 1830's, thousands of the tribesmen refused the oath of allegiance to the new French governors and preferred exile. Their leader was the noble Abd-el-Kader, famed in story and song, on the subject of whose heroism and gallantry my father was wont to recite an immensely long and tedious poem. The Turks gave them asylum in what is now Trans-Jordan, and there they have lived in strictly segregated communes, ruled by descendants of their tribal hero. They are engaged in the bitter strife with all their neighbors, especially with the Bedouin who roam the desert east of their settlements.

Farther south again, in Biblical Edom, where the sun hung over our heads like an immense copper caldron that sent out boiling vapors, opposite the city of Jericho, east of the lower reaches of the Jordan River and the Dead Sea, the Turkish government planted, towards the close of the last century, a long string of settlements for Circassian immigrants from the Russian Caucasus. Their descendants, easily recognizable by their platinum blond hair and blue eyes (their women were the favorites in the harem of Abdul Hamid), are at present subjects of Emir Abdullah of Trans-Jordan, who is the late Sherif Hussein's son and a brother of the late king Feisal of Iraq. Abdullah is the perennial stand-by of journalists paying a quick visit to the Holy Land who want to be able to say that they have seen at least one Arab prince whose appearance and environment corresponds to Hollywood's conception of a grand sheik.

Abdullah, who is fat and lazy, but shrewd, fox-visaged when off his guard, and well informed on world affairs, will receive any visitor who breaks the monotony of royal existence, under which he suffers keenly. He told me, a few years after the British had installed him in his insufferably dusty little capital of Amman, that he is bored to death with the Arabs, the British, the Jews, his wives, his palace, his plush furniture, and chess and that his only solace lies in aspirin and an occasional swift ride in his Rolls-Royce across the Allenby Bridge into Palestine. He expressed himself in terms of admiration for Winston Churchill. And no wonder! It was Churchill who, by carving him a good-sized though empty kingdom out of Palestine's eastern half (which consists chiefly of grazing grounds), rescued Abdullah from the desuetude into which his father's fortunes fell after the Sherifian army was defeated by Ibn Sa'ud.

Like all Arab potentates, Abdullah is for an Arabian Empire—provided he, and not Ibn Sa'ud, is made emperor. For the rest he waits, plays checkers, and hopes for the best. Unlike his late brother, Feisal of Iraq, who was reconciled with King Ibn Sa'ud after that prince expelled their father and brother Ali from Mecca and drove them into exile, Abdullah cannot forgive. Too weak, however, to take revenge, he sticks closely to the British in the hope that his loyalty will some day be rewarded with something more substantial; a throne in Jerusalem or Damascus, for instance. "All things are possible to him who believeth," said the Emir. "My brother lost Damascus and won Baghdad. If Allah wills and I live, I may lose Amman some day and I shall not be sorry."

Trans-Jordan is sparsely settled. There are considerably less than half a million people in the emirate. Half of these are Bedouin who roam about in the hinterland, near the borders of Ibn Sa'ud's domains. Nothing has been done to improve their lot or the agriculture of the country since the Turks left, and there are no industries, although we found that there has been a trickle of immigration from Tripoli, in north Africa, since Abdullah mounted the throne. The immigrants are Senussi who live in small settlements along the old pilgrim road. They left their native land after Rodolfo Graziani, the man who was also sent to "pacify" Ethiopia, had their most venerable patriarchs taken up in airplanes and thrown to their death on the rocks of the Halfaya Desert. Except for a few thousands who found asylum in Egypt, the Trans-Jordanian Senussi are probably the only survivors of their tribe, for during Graziani's ten years' governorship the native population of Tripoli, according to a report issued by the League of Nations in 1934, was reduced from two millions and a half to sixty thousand. The Senussi in Trans-Jordan take no interest in Arab affairs, for they consider themselves temporary sojourners who plan to return to their homeland as soon as and if ever conditions warrant.

Although the gaps in the Hejaz Railway that Lawrence blew in it had been repaired, we noticed that the station houses and the former Turkish military posts and blockhouses throughout Trans-Jordan lay deserted and overgrown with weeds. From Abdullah's domains southward, in the regions adjoining the Red Sea, the population is homogeneously Arab but split up into semi-independent tribes: Ageyl, Serahin, Ruallah, Beni Sakir, Howeitat, and scores of other, smaller ones who maintain a jealous isolation and separateness that no amount of preaching by the propagandists of Arab nationalism has been able to overcome. Deep in the interior, protected by howling deserts, unreachable, except by armed caravans, are more independent tribes, powerful, said to be warlike, defiant of Ibn Sa'ud and the British, ruled by the irascible Sultan Ibn Rashid, Imam of Sehr, and the potentate of Oman. Lava fields, gravel fields, sluggish wadis with a few anemic palm bushes, sand dunes, sand oceans immeasurable, and here and there a sloping hill on which vast herds of goats nibble at invisible grass are predominant in the scenery of lower Trans-Jordan. For the rest, vast stretches of silence, immobility, and repose, vast spaces reserved to instinct, to the half-sleep of thought where life moves to the rhythm of tribal quarrels with their raids and sudden death, as in the world's oldest times.

The only cultural link between the tribes is their language and religion. The Arabs are the world's original and most determined home rulers. Five families make up a nation. They have no conception of a common task or a common future. Life is disfigured by a universal poverty of unimaginable ugliness. Everywhere in the settlements we saw the same unkempt, black-smocked women around the wells and hordes of naked children; the same seemingly aimless and hopeless existence. We ate their meals occasionally, at the risk of being covered with lice and fleas from the sheepskins and rugs in their tents: boiled mutton in rice was the greatest feast the wealthiest could offer. Among the humbler people the staple food is dates and unleavened bread, both scarce in the year we traveled about because of the drought and a bad harvest. In some villages the inhabitants, emaciated by starvation, were sitting quietly in the doorways of their huts waiting for death, yet without despair or protest. *Inshallah!* If God wills, it must be! To speak with these people of empire, of federation, of political ideas, is waste of breath and puerile nonsense, unless it is dished up with a prospect of riches and loot. Pan-Arab ideas are in fact confined to the heated centers of intrigue-Damascus, Baghdad, Jidda. Elsewhere the spirit of union is being grafted on the discontent generated by landlord evictions and long-standing injustices to the peasants and nomads. Poverty and hopelessness make Arabia a great inert mass that can be swayed now this way, now that by the agents of a foreign imperialism.

Still, in all those peoples and tribes of the Peninsula there lives the remembrance of a common past glory, which is kept alive by storied song and epic poems dealing with the deeds and splendor of the great khalifs. Six hundred years of Turkish oppression has not been able to obliterate that memory. But, as is the case with all conquered peoples whose golden age lies in the past, the triumphs of long ago are exaggerated to fantastic proportions. Legends have become interwoven with historical narrative. In the sordid present the past has become a refuge, splendid almost beyond human conception. There, amid the sandy solitude, the grim rocks, and the boundless emptiness of the desert where length, breadth, and height and time and place are lost, and nothing but his own questioning voice breaks the stillness, the Arab broods over what his ancestors achieved in the past. This perennial introspective mood, Islam folded back upon itself and orientated towards secrecy, gives its men a sense of disdainful superiority over the

Westerner they occasionally encounter. Nothing that we have, nothing that the Turk or Britain or France represents or can give him, compares with what was once accomplished in the *gesta Dei per Arabos.*

And indeed great was their work: to cut a path for the truth in the world, the believers must in doing so hurl themselves against the mightiest empires and face odds as fierce as the thunderstorms that rage in the waste places of their native habitat. One single motive impelled them, one idea drove them on: to proclaim and enforce the theocracy of Islam. Unlike Tamarlane or Alexander, they had no eye or heart for the things of this world and utterly contemned the riches of the West, of whose existence, in fact, they were ignorant. With them, as with Cromwell and the armies of the Saints in Bohemia and Holland, it was hunger for an assurance that the forces of the universe are on the side of the elect. They were a chosen people, conscious of their destiny in the providential scheme of things and determined to translate it into reality. What may look, across the distance of thirteen centuries, like an unintelligible whirlpool or a volcano blindly throwing up its flames and ashes, became the cause of immeasurable change for the peoples of all the earth. Mohammed and the khalifs set a chain of events in motion that to this day influence the acts of man in Europe and America.

In a few years the Arab armies swept around the Mediterranean and through western Europe to the very gates of Paris, enveloping Christendom in an embrace that came near to being fatal. The Crusades did not arrest the march of Islam. When the Crusades got under way the tide had begun to recede. The waters were stagnating in half a dozen places. The magnitude of the triumph burned out the zeal of the armed missionaries. The conquerors fell on their own swords. The central idea, which was the propagation of the faith, lost its all-compelling power. Sometimes, when initial reverses overtook it, it looked as if Islam had merely *reculé pour mieux sauter,* but a fratricidal struggle between the successors and epigoni of the great khalifs gnawed at the root of the movement and caused the branches to wither.

Thereafter Arab history can be seen only in the framework of rival dynasties in the east and west, in the flowering of civilizations in Egypt, Morocco, Iran, and Spain, in the rising Ottoman Empire, and finally in relation to the Eastern question—that is to say, as a function in the modern European balance of power. Arab kingdoms, protectorates, fiefs, and other local institutions are felt as something parochial and subordinate. From servant and equal in the ranks of those confessing Allah, the Osmanli Turk became the master of all. After that, Arab society, as Lawrence says, was a subsociety, or rather a chain of subsocieties stretching from the Indian to the Atlantic Ocean, split asunder by the claims of rival caliphs, religious schisms, and commercial rivalry. There has never been a distinctive and satisfying national culture that was common to all the lands of Islam and the peoples originally brought under its sway and upon whom it imposed the Arabic tongue. Egypt on one end of the Mediterranean and Morocco on the other, Iraq and Iran on the Persian Gulf, and beyond these, in the still almost inaccessible

East, Bokhara and Samarkand were the most desirable provinces in the heritage of the primitive Arab conquerors. But they fell under foreign domination or independently pursued their way.

In the Peninsula, especially in the southern part, the real Arabia, "the only part of Arabia that matters," as Philby would have it, life has long since reverted to the pattern of antiquity. Its clashes and wars are scarcely noticed in the world at large. One would be almost justified in saying that in the heart of Arabia the rhythm of Canaanite times has returned with its provocations and shocks occasioned by tribes, driven by hunger, impinging upon their neighbors and repeating the eternal process that was under way even when the Israelite wanderers came up from the desert, exchanging the nomadic for a sedentary system of economy.

My chief impression, next to the vastness of the territory inhabited by Arabs, was the wide divergency and apparent irreconcilability of the tribes. The feeling, which was renewed each of the seven times I returned, was one of overwhelming sadness and terror. Arabia, not the cosmopolitan Syria or modern Palestine, but the Arabia of Trans-Jordan, Iraq, Oman, and the hinterland of Aden and Saudia, is terrible. As I wandered to and fro, looking on, asking questions and feeling my way, it was almost inconceivable to me that once these lands were bubbling caldrons where life seethed and steamed and boiled over, that here fertile minds, both receptive and creative, were imbued with ideas that propelled the forward march of mankind. If it were not that they have left their monuments as far as Spain and Tashkent, if the imprint of Islam were not visible from Damascus to Peking, I would almost believe that the waves raised by Mohammed and the khalifs are as much a figment of the imagination as *The Thousand and One Nights,* enchanting no doubt, but a dream, a desert phantasmagoria of opulence and living water.

Was it the Turk? Was it the more or less impure misalliance with the bureaucratic Osmanli that withered Arab creativeness and dried up their imagination? Whatever the cause, it is as if a blight has passed over this land and its people, as if a curse rests upon it. Everything is trivial and ephemeral in present-day Arab life. There seems to be nothing to encourage the men to greater effort. The tribes vegetate in physical misery, without any apparent hope of rising out of the slough. It is pleasant enough to go on expeditions to Arabia Felix, to ramble and to explore, to see the dark, scorn-speaking countenances of the tribesmen and try to bring at least to an individual here and there the divine relief of communication. Arabia is terrible and yet it attracts one almost magically.

What was it then that drew me again and again to those sterile, sun-crushed rocks, the burning bush of tamarisk and cactus? Was it not perhaps the intoxicating silence of those soundless nights when the earth is but a pale shadow about your feet and above your head is "heaven opening into utter heaven, boundless height upon height with stars, with stars, with stars," that come to you as lovers gently swaying, stooping, ringing tiny silver bells that murmur the prelude of morning? On those nights the heavenly bodies appeared as ships rolling upon seas without shore, worlds dancing on the ruins of time, almost within reach of your finger tips. There alone, in that austere solitude, you become aware of why those mad, God-struck Semites—Moses, Jesus, John the Baptist and Mohammed—sought the desert: to be alone, to be earth-exiled, to seek a home for the mind in the bloom of midnight, to hear the phantom wings of eternity, the beating of God's heart! In that stillness, with only the lights in the Lord's mansion illumining his path, man becomes a creature of infinite flight, and time flies like a furtive shadow.

I cannot count the evenings when I rode aimlessly not knowing what I was doing at all in a land where the sun is a torture to me, feeling the immense sadness of that world. Perhaps it was one of those sadnesses, strange as it may sound, in which life fulfills itself just when one thinks to lose it, when we no longer care what we are or whence we came but just feel ourselves brothers to all mankind and when we would call out: men, take stock of your condition! You have a right to full human existence. Fight yourselves free! Break the chains that bind you to feudalism and to the myths of the past. The golden age lies before you, not behind! . . . .

CHAPTER 3

BRITAIN'S ROLE IN PALESTINE

A plain-spoken allegation, which has gained wide currency of recent years in America, would have it that the chronically disturbed or uncertain condition of Arab-Jewish relations in Palestine is at bottom the result of an early piece of unconscious or deliberate muddling on Britain's part. It has almost come to be taken for granted that all the trouble in the Near East goes back to the First World War, when, it is said, the British government allowed one of its departments, the Foreign Office, to promise Palestine to the Jews, while another department, disregarding the more or less solemn pledge to world Jewry, or oblivious to its significance and scope, almost simultaneously entered into negotiations with the Arabs with a view to including the Holy Land, on their demand, in an eventual independent Arab state or federation of states. So ambiguous and dishonorable a procedure, which some are quite ready, without seeking further substantiation or historical confirmation, to impute to the British government, has led to a widely current notion that Palestine is a "twice-promised land" or even "the too-much-promised land."

This conception, it cannot be said too emphatically, rests upon a totally erroneous interpretation of historical fact. To Great Britain attaches no guilt of double-dealing or deception in the matter of having made mutually conflicting promises to Jews and Arabs. Britain did indeed make important promises to both these peoples. But these promises did not conflict or mutually exclude one another. Moreover, the promises made to the Arabs were nearly all carried out in spite of enormous difficulties involved in their fulfillment.

On the other hand, insofar as Palestine is concerned, that country was promised only once and to one people only: the Jews.

In the oral and written promises made to the peoples inhabiting the huge Arabian quadrangle as inducements to bring them as active participants into the war against Turkey, Palestine and its future were left wholly out of consideration. And this for a very good and valid reason: It was neither by oversight nor was it a calculated omission that Palestine was not even once mentioned by name in the various plans and drafts which were proposed and discussed by representatives of the Great Powers in the course of the First World War for the reapportioning and the reorganization of the erstwhile Turkish domains in the Arabian peninsula: Palestine simply did not count. It offered no possibilities of bargaining one way or the other.

None of the great Arab chiefs, from Hussein to Feisal and Ibn Sa'ud, and none of the Arab revolutionary committees, such as those of Syria and Mesopotamia, which plotted the overthrow of the Ottoman power, as much as introduced the subject of Palestine in the negotiations, both secret and open, that were held with British diplomatic agents. Nobody gave Palestine a moment's thought. What is more: not a single Palestinian Arab, either lord or commoner, appeared at Feisal's headquarters or at the British agency in Cairo, in Baghdad or in Jidda, where negotiations between Britons and Arabs, at one time or another, were in progress during the better part of three years.

Even after the issuance in November, 1917, of the Balfour Declaration, which was by no means a secret document and which was universally interpreted at the time as containing Britain's pledge to shape political conditions in such a way that a Jewish commonwealth might come into existence in Palestine after the war, and of which the Arabs were fully aware, not a word of protest was voiced against the project by any Arabs, either in Palestine or outside.

No Arab began to protest against the Jewish settlement of Palestine until agents of the British Colonial Office—that is to say, members of that department of the British government which had conceived the scheme of a quasi-independent Arab federation—began suggestively, to express their regret that Palestine was not included in their plan but had been set aside for the Jews.

Palestine was a wilderness before the war, inhabited by a few hundred thousand poverty-stricken fellahin, the most apathetic to the prospect of freedom in the whole Arabian Peninsula and culturally the most backward. The Palestinian Arabs were not stirred in the least by the appeals emanating from the Arab chieftains and British agents to help throw off the Turkish yoke.

Not the smallest of their villages rose at the call of freedom. Most of the prominent Palestinian Arab families remained loyally on the side of Turkey during the entire course of the war. For instance, that scion of the noble family of the Husseinis who was to become Mufti of Jerusalem after the British occupation of the city who instigated the anti-Jewish excesses in 1919 and 1920 and who twenty years later, after a career of intrigue and bloodshed, had a hand in fomenting the Iraq rebellion, served as a volunteer in the Ottoman army. Palestine cut no ice militarily, economically, or culturally in 1917.

Palestine did, of course, contain the Holy Places, but on the subject of these, as Father Joseph Lagrange of Jerusalem, the great Dominican exegete, who remained in the city during the entire course of the war, informed me, "No uneasiness was apparently felt by anyone, except myself." Dr. Lagrange was afraid that, as a result of the first secret plans to partition the Turkish Empire in which Russia had a hand, the custody of the Holy Places might fall to the ecclesiastical representatives of Orthodoxy. The October Revolution, which occurred a month before the issuance of the Balfour Declaration, he said, put an end to his fears once and for all.

At the start of the First World War the only more or less definitely envisaged change in the Near East concerned Syria. This country, it was tentatively agreed between Britain and France, with Russia somewhat reluctantly consenting, was to pass into the hands of France. Only the Syrian patriotic societies and the secret revolutionary committees in the Syrian cities were kept in complete ignorance about the understanding at which the Great Powers had provisionally arrived. Had they been aware that it was intended to pass their country from Turkish suzerainty to French control, they might well have denounced the plan and have become extremely troublesome. That they seriously suspected France of having designs on their country is evident from the frequent representations made to Great Britain by Syrian agents and the assurance they asked that not France but Britain herself undertake the liberation of Syria and, after the expulsion of the Turk, institute such political conditions as would speedily lead to self-government and independence.

A second drawback to an unambiguous understanding between the parties interested in the defeat of Turkey was the fact that France, which had been promised Syria in secrecy, understood that country to include both Lebanon and Palestine. This is quite understandable when it is borne in mind that under the Turks Palestine was administered from Beyrouth. It formed part of the vilayet of Beyrouth, the administrative bounds of which extended as far south as Gaza, in Palestine. The French therefore took it for granted that, in the event of their obtaining control of Syria, they would automatically succeed Turkey in all of Syria's constituent provinces and administrative departments—in other words, they would rule in Damascus, capital of Syria, Beyrouth, capital of Lebanon, and Jerusalem, capital of Palestine. To this French pretension the British countered with a repetition of the reminder previously addressed to the Russian Government that *"Palestine must be reckoned a country whose destiny must be the subject of special negotiations in which belligerents and neutrals are alike interested."*

France accepted this interpretation with good grace and before the end of the war *surrendered her interests in Lower Syria, i.e., in Palestine to Great Britain on the distinct understanding that it was to be placed under international control and that a Jewish national home was to be established in the country in accordance with the terms of the Balfour Declaration.*

The contention of present-day Arab nationalists and their sympathizers that the country of Palestine was "from the beginning" understood to lie within the boundary of such Turkish territories as it was hoped and planned to liberate from the Ottoman oppressor, and "therefore," automatically as it were, included in the congregation of independent Arab states to be created after the war, does not stand up in the light of Anglo-Arab treaties, understandings, protocols, and the official correspondence exchanged between the high contracting parties. For while it was unquestionably intended to drive the Turk out of the Holy Land and it is also certain that Palestine's liberation did not come about accidentally or rather incidentally to a general British military campaign in the Middle East, but as a result of military operations specifically aimed at the occupation of that country, the matter of Palestine's future disposal was clearly reserved, and to the knowledge of all concerned, for special international deliberation and disposal.

Even before the Balfour Declaration was issued the Arab leaders knew, as well as everybody else, that Palestine's future allocation was linked with a project to transform it into a national home for the Jewish people. Palestine's conquest and the decisions regarding its eventual status did not come about as something unforeseen or unpredictable in the course of military operations and no hasty makeshift arrangement had to be made regarding it and its future after the battles of Gaza and Megiddo. The conquest of Palestine was one of the main political objectives in the British war against Turkey. It was planned in minute detail also as regards its future reorganization. Yet for years the Arab nationalists and these Arab princes who were theoretically on Britain's side in the war never by so much as a word objected to or protested against the contemplated Jewish settlement of the Holy Land.

Upon the selection of Hussein, Sherif of Mecca, as their spokesman by the Arab nationalist leaders and their demand that the irascible old potentate's terms be met, negotiations were finally begun in earnest with the British in July, 1917. The conditions on which the Arabs were prepared to come into the war against Turkey were formulated in the so-called Damascus Protocol which Britain was given thirty days to accept or reject. The Arab princes demanded recognition by Great Britain of the independence of the Arab countries lying within the following frontiers: north—the line Mersin-Adana to parallel 37 N. and thence along the line Birejik-Urfa-Mardin-Midiat-Jazirat-Amadia to the Persian frontier; east—the Persian frontier down to the Persian Gulf; south—the Indian Ocean (with the exclusion of Aden, whose status was to be maintained), and west—the Red Sea and the Mediterranean Sea back to Mersin.

The British replied a month later, through Sir Henry McMahon, British High Commissioner in Egypt, that they accepted the Arab terms in so far these were compatible with French interests in Syria. This meant, of course, a definite exclusion of Palestine also, since that country was then administratively a part of Syria. With regard to the frontiers, Sir Henry remarked that "it would appear premature to consume our time in the discussion of such details."

Although the British government, in view of the disastrous military situation of the Allies, was most anxious to bring the Arabs to revolt against Turkey and thus relieve Turkish pressure on the Russian armies in the Caucasus, it looked upon Hussein's demands as exorbitant and bordering on the fantastic. London was well aware of the fact that the Sherif's small prestige in the Arab world had sunk still lower since it had become known that he aimed at the re-establishment of the khalifate, which had been abolished by the Young Turks, in his own bailiwick of Mecca, and was quietly hoping to reserve for himself the dignity of Khalif ul-Islam, or Commander of the Faithful, formerly the prerogative of the Turkish sultans. Now, to the Shiite Moslems—those of Mesopotamia, for instance—Hussein was a Sunnite heretic and his Hejaz tribesmen a mob of uncouth barbarians.

Ibn Sa'ud, Emir of the Ṇejd, detested Hussein and in the course of time was to chase him from the throne Britain set up for him. Hussein was furthermore known to have no influence whatever in Syria. The revolutionary nationalist leaguers in that country did respect his third son, Feisal, the future king of Iraq, who had served in Syria on the staff of the Turkish governor-general. But in the narrow-minded and selfish old Sherif himself they had no confidence. The British knew that Hussein, in spite of all his pretentions, could not speak for Syria. Even so, because of the gravity of the military situation, they accepted his demands as a basis for negotiations, with a reservation on the subject of France's claims in Syria—that is, the present countries of Syria, Lebanon, Palestine, and Trans-Jordan.

In his reply of September 9, 1915, Hussein strongly objected to the evasive character of the British note on the subject of frontiers. He demanded a definite answer and an immediate decision on all future boundaries, threatening to break off the negotiations if Britain did not react favorably forthwith. Before responding to the Sherif's peremptory epistle, which was accompanied by a violent printed denunciation of British perfidy which had appeared in the official *Gazette of the Hejaz* in the form of an editorial written by Hussein himself, the British agency in Cairo established contact with the Syrian nationalists for whom, among others, the Sherif pretended to speak. From them it was learned that they would rise in revolt for the liberation of the districts of Damascus, Aleppo, Homs, and Hama, but that they would respect Britain's reservations with regard to the areas in which other powers were interested. This, obviously, referred again to the vilayet of Beyrouth, or the coastal regions of Syria—that is to say, Lebanon and Lower Syria, as Palestine was then known in official language.

The contemplated revolt in Syria, however, never materialized. The Turkish secret police got wind of the plot, arrested all the nationalist leaders, and executed them to a man. Thereafter, and for the duration of the war, Jemal Pasha, the Governor-General, instituted so rigorous a system of political control that not the faintest glimmer of revolt was discernible till the day he departed, which was also the day General Allenby entered Damascus.

Sherif Hussein, on his end of the Peninsula, also showed little zeal in starting the revolt. Again and again he postponed the issuance of a proclamation announcing his break with Turkey, and his uprising would probably never have taken place either had not the Germans forced him to make up his mind.

The German High Command sent a military mission to southern Arabia to investigate the rumored disaffection of various Arab tribes. The Turkish government supported the mission with two divisions of infantry, a brigade of artillery, and a cavalry force. After what had happened to the Syrian nationalists, the old Sherif could have little doubt of what fate was in store for him should the Turkish force arrive in his regions. He therefore broke the long suspense by proclaiming the independence of the Hejaz, overwhelmed the small Turkish garrison in Mecca, and attacked Medina. The Turks in Medina threw his crowd of Bedouin back with losses so serious that the attack was never renewed. And the Turkish commander in Medina held out until the Peace Conference notified him that the war was over and that Turkey had ceded the city and the rest of Arabia.

Even so, that Arabs were in the war at last. When they did come in there had been no change in the conditions upon which they had made their collaboration contingent—that is to say, the conditions granted them in Sir Henry McMahon's letter. Britain was to recognize the independence of the Arab countries after the war but reserved her decision in the case of the Syrian coastal region in which France claimed an interest. With this understanding in the background, the British proceeded with their two great parallel military campaigns: the one in Mesopotamia with Baghdad as the ultimate objective; the other in Sinai and Palestine with Damascus as the final goal.

When in later years an Arab nationalist movement was born in Palestine and its leaders, in support of their opposition to the Jewish national home, invoked the correspondence that had passed between Sir Henry McMahon and the Sherif of Mecca at the start of the war and incited the peasants to bloodshed and destruction of property by alleging that Britain had broken her pledge to make Palestine an independent Arab country, Sir Henry McMahon, then long since retired from government service, wrote two letters to clarify the earlier understandings. In the first, dated March 12, 1922, addressed to the British government, he said that he had intended to exclude Palestine from the area of Arab independence as fully as the Syrian coastal regions to the north. In a second letter, addressed to the editor of *The Times* and published by that newspaper on July 23, 1937, he wrote: "I feel it my duty to state, and do so definitely and emphatically, that it was not intended by me in giving this pledge (of independence) to King Hussein to include Palestine in the area in which Arab independence was promised.

I also had every reason to believe at the time that the fact that Palestine was not included in my pledge was well understood by King Hussein."

Sir Ronald Storrs, who as Oriental Secretary to the High Commissioner (Sir Henry McMahon) handled the Anglo-Arab correspondence, wrote in his *Orientations*: "Palestine was excluded from the promises made to the Arabs before those British (military) operations which gave freedom to so large a proportion of the Arab peoples."

In the year 1937, when Palestine was plunged into chaos and bloodshed by the Arabs, and their leader, Haj Hussein Amin, Mufti of Jerusalem, again invoked the McMahon correspondence, William Ormsby Gore, who was attached to the McMahon staff in 1916, stated in the House of Commons, on July 21, 1937, "that it never was in the mind of anyone on that staff that Palestine west of the Jordan was in the area within which the British Government then undertook to further the cause of Arab independence." Colonel Lawrence also took the view that Palestine was excluded from the areas in which Britain intended to foster Arab independence and said so in a letter to *The Times* on September 11, 1919.

Italy's entry into the war in 1915 on the side of the Allies, an act that the Rome government had made conditional on obtaining a share, among other things, in the loot expected to result from the defeat and the elimination of the Ottoman Empire from international affairs, required a reconsideration of the scheme of dismemberment provisionally agreed upon by Britain,

France, and the Arabs. The apportioning to Italy of the port-city of Smyrna and its hinterland, plus a recognition of her claims to interest, in southern Anatolia, which is Turkey proper, did not appear to affect the Arab position even remotely. Hence the advent of Italy into the Allied camp passed without a ripple at the time.

Different it was when the Bolsheviks, in October of the following year, having seized power, proceeded to open the secret archives in the Foreign Ministry in Petrograd and found there a copy of the Sykes-Picot Agreement. This document dated from the fall of 1915 and dealt with the future of Asiatic Turkey. It was drawn up by Sir Mark Sykes for Britain and Henri Picot for France. Of the existence of this treaty the Arabs seem to have been kept in total ignorance. When Leon Trotsky published its terms and the Arabs learned its contents from the German and Turkish press, they were thrown into consternation. They were not so much alarmed by the newly discovered document's provisions as by the fact that they had not been apprised of its existence.

They suspected double-dealing. They wanted to know what other treaties there were. On which agreement were the Arabs to place reliance? Were there by chance any other skeletons in Allied cupboards which another historical accident might bring to light? The British agency in Egypt sought to calm the storm of indignation in the Arab camp by referring to the British pledges contained in the McMahon correspondence. But Feisal asked Lawrence on which of the two his father, the Sherif, was to rely. To this the Englishman, who was more irritated than anyone else, replied that it would be wisest always to take the last word for the final decision. The trouble was precisely to discover which was Britain's last word, for the Sykes-Picot Agreement antedated the McMahon correspondence, although it had but latterly come to the knowledge of the Arabs.

Fundamentally, however, there was no difference between the two so far as Arab interests were concerned. Britain promised independence to the Arab countries in the Sykes-Picot Agreement as she had done in the McMahon correspondence, and she excepted Syria for France as she had done before. The only new element which came to light by the publication of the Sykes-Picot document was Russia's inclusion amongst the beneficiaries of the Turkish Empire's dismemberment. Russia had been assigned a large slice of Turkey's Caucasian provinces.

But that again could scarcely be said to interest the Arabs, for whether Russia installed herself eventually in Turkey's Armenian and Kurdish provinces or not would in no way interfere with a fulfillment of Arab national aspirations. Moreover, when the Bolsheviks made public the terms of the Sykes-Picot Agreement, it had virtually been nullified by military and political events in so far as it concerned Russia's participation in the dismemberment of the Turkish Empire. The October Revolution had taken Russia out of the war, leaving czarist ambitions both in Thrace and in the Caucasus unrealized. The Bolsheviks showed little in clination to press these or any other territorial claims of their imperialist predecessors.

The storm in the Arab camp soon blew over. Peace

came and with it the struggle for Arab independence entered a new phase. The armistice with Turkey found the British armies victoriously installed in Jerusalem, Baghdad, and Damascus. Temporary military administrations were set up in all three of the capitals, awaiting a permanent political settlement of Arab affairs by the peace conference. But the peace conference, it soon became evident, would for a considerable time delay the final accounting with Turkey. In fact the peace treaty with Turkey was not signed till five years later, in 1923. Many other, more pressing problems demanded the immediate attention of the Allied statesmen and diplomats gathered in Paris. The delay and the prospect of still more delay caused the Arabs to grow restless. Seven of their most eminent leaders journeyed to Cairo and expressed their apprehensions at the British agency that the delay might lead to permanent military occupation of the Arab countries. In order to set Arab fears on this score at rest, the Allied Powers caused a joint declaration to be published throughout the Arab world in November, 1918. This document, known as the Declaration to the Seven, reads as follows:

"The object aimed at by France and Great Britain in prosecuting in the East the war let loose by the ambition of Germany is the complete and definite emancipation of the peoples so long oppressed by the Turks and the establishment of national governments and administrations deriving their authority from the initiative and free choice of the indigenous populations. In order to carry out these intentions France and Great Britain are at one in encouraging the establishment of indigenous governments and administrations in Syria and Mesopotamia, now liberated by the Allies, and in the territories the liberation of which they are now engaged in securing, and recognizing these as soon as they are actually established. Far from wishing to impose on the population of these regions any particular institutions, they are only concerned to ensure by their support and by adequate assistance the regular working of governments and administrations freely chosen by the populations themselves. To secure impartial and equal justice for all, to facilitate the economic development of the country by inspiring and encouraging local initiative, to favor the diffusion of education, to put an end to dissensions that have too long been taken advantage of by Turkish police, such is the policy the two Allied Governments uphold in the liberated territories."

No specific reference to Palestine was made in this Declaration. But then it was not a new statement of policy, but rather a reiteration of principles and of former pledges and war aims by the Allies. The purpose of the Declaration to the Seven was to reassure the Arabs that there was really no change in Britain and France's intentions with respect to the future of the Arab lands. It did not replace former agreements and declarations of policy. France's claims to Syria were by no means abandoned under the Declaration, except that she had transferred her rights to Lower Syria (Palestine) to Britain.

The Arabs accepted the document as satisfactory. No protest was raised against it anywhere. This meant that they had also accepted not only France's reservations on the subject of the Syrian coastal regions, but also the transfer of these French rights in so far they concerned Lower Syria (Palestine) to Britain for the purpose of establishing the Jewish national home there. The Balfour Declaration had been issued. The transfer of French rights had been a public transaction. The Arabs were aware of both and yet they raised no objection.

However, opposition of a most violent kind to the political reorganization of Syria by France and to the settlement of Jews in the Holy Land was soon to manifest itself in a most tragic manner. But that opposition, strange as it may seem, did not spring up as a spontaneous reaction in Arab nationalist circles, although Arabs were subsequently to give expression to it in word and deed. It originated and was carefully nurtured in the milieus of the newly arrived British civil and military administrators. Most of these men had been hastily drawn from the British civil and military services in Egypt. Some came from the Sudan, some from Kenya and Rhodesia, others from India. They were nearly all juniors in the colonial service or had held subaltern positions in the army. None of course had any experience in the administration of a territory that was marked for some sort of a special international dispensation as was involved in the term *mandate* and as distinguished from the old colonial system in which the relationship of governors and governed is simply that of overlords and natives.

Nor were they all of the high moral and intellectual caliber of the old-type British colonial functionary. In fact, some of them with whom I was personally acquainted fell far short of an elementary public-school education. One man in particular who, immediately upon his arrival in Palestine, was placed in a high administrative position, in recognition of some conspicuous act of bravery performed on the Mesopotamian battlefield, had been a butcher's assistant in what corresponds to the Commissariat Department in Cairo till 1917. He could scarcely read or write. Yet a large, densely populated district was placed under his control. In conversation he revealed himself an inveterate and vulgar xenophobe, expressing himself in terms of contempt and hate for everything and everybody not British. That he deigned to speak to me at all in his new exalted position, I owed, no doubt, to the fact that he knew me to be a British subject. He knew neither Hebrew, Arab, nor French and disdainfully rejected suggestions to familiarize himself somewhat with the history of the two Semitic peoples over whom he was to rule or acquaint himself with the rudiments of Turkish law which remained valid in Palestine.

Another British functionary, Sir Ronald Storrs, the first Governor of Jerusalem, on the other hand, was a fanatical *arabisant*. He idealized everything Arabic, much like those writers about foreign folkways who idealize the peasants no matter how backward in culture, crude in manners, and unappetizing in appearance they may be. Storrs spoke all the languages of the Near East. With his amazing versatility, he combined an affectation of scholarship and a mask of geniality to hide a haughty condescension for lesser breeds such as Frenchmen and Jews. That man had a great deal to do with the shaping of British policy in the early days of the British occupation of Palestine, when the building of the Jewish national home got off to an uncertain start. He more than anyone else was responsible for laying the foundation of that anti-Zionist policy of the successive British administrations in Palestine. And Storrs acquitted himself so well in the early years that in 1943, when the liquidation of the Jewish national home was stealthily placed on the program by the Colonial Office, he was recalled to Palestine from Northern Rhodesia, where he had filled a term as Governor, to bring his interrupted task in the Holy Land to fruition.

Between these men—Ronald Storrs, Louis Bols, Harry Luke, John Chancellor, Humphrey Bowman, Harry Keith-Roach, even Herbert Samuel, the first High Commissioner in Palestine, and that extraordinary woman Gertrude Bell, the British agent who went up and down the Near East, as Lawrence did, long before the war with Turkey, to spy out the lay of the land—between all those people, the chief actors in the Palestine drama, there grew up a remarkable community of spirit on the subject of the building of the Jewish national home. They constituted themselves into a more or less secret brotherhood or society of watchdogs. They had no instructions. No official declaration had ever been made on an eventual Jewish majority. Yet they early decided for themselves that their main task was to see to it that such a majority would never be attained. They were the genuine imperialists, for their opposition to the eventual erection of a free and independent Jewish Commonwealth was predicated upon fully warranted apprehensions that the freedom of one people, the Jews in this instance, in the traditional area of colonialism would in the course of time, as a small flame lighting a huge pile of wood, set the whole colonial world afire and ring down the curtain on the imperialist episode of usurpation and spoliation. In their anti-Zionist policy they were loyal to the supreme interests of British imperialism.

I include Sir Herbert Samuel in this company of the select, although he is a Jew. His appointment as the first High Commissioner in Palestine was one of the shrewdest moves pulled off by the imperialists of the Colonial Office. With their deep insight into the character of assimilationist Jews, the fabricators of the policy of Empire surmised that such a Jew as Sir Herbert would go out of his way, in order to reward the confidence placed in him by being supraimpartial, extraneutral, and 125% loyal to British interests, which in this case meant slowing the building of the house for his own people. They were not wrong in that surmise. Sir Herbert Samuel performed according to unwritten specifications.

Humphrey Bowman, now retired from the colonial service, reveals in his book of reminiscences* how he wrote a letter to Gertrude Bell, who was stationed at Cairo in 1921, in criticism of the Balfour Declaration. Himself he was then, after having served in two Arab countries, in charge of the Department of Education in Palestine. Miss Bell wrote back: "The long letter I had from you (on conditions in Palestine) gave me just the impression I expected . . . . However hard may be our task in Mesopotamia, at least we (the British)

**Middle East Window*, Longmans, Green, and Co., Ltd.

are floating down whatever current there is, the stream of nationalist sentiment, which is after all the only visible movement." Miss Bell then goes on to foreshadow what she regards as "the inevitable failure of the Zionist effort," provided the Arabs proceed along their path patiently: "The Arabs have enough to do, if they will only believe it, in organizing what they have got, and if they will be patient (which they never will!) whatever else they deserve will drop into their lap, and not all the gold of the Hebrews in the world can prevent it."

This is from a correspondence dating from 1921, exchanged between two persons who belonged to the coterie of British colonial servants charged with the supervision of the execution of the British mandate over Palestine and of the terms of the Balfour Declaration: "His Majesty's Government looks with favor on the establishment of a national home for the Jewish people in Palestine . . . facilitate the close settlement of Jews on the land . . . ."

"The Hebrew gold!" That point of view seems to have covered not only the mind of Gertrude Bell, who exerted great influence in Near and Middle Eastern affairs, but also many a subordinate official's approach to the task in hand. Bowman himself states that the Jews in Palestine, because of their funds, were able to take care of their own education, while the government, naturally, had to establish schools for the Arabs throughout the country, giving only a per capital grant to the Jews . . . . What he does not say is that the per capital grant came and comes from the general taxes of which the Jewish minority pays 85% of the total and that the Jews thus, indirectly, virtually take care of Arab education, too.

The Hebrew gold in Palestine consists of the pennies of the Jewish poor and the Jewish petty bourgeoisie all over the world. The rich Jews have never been interested in their Palestinian homeland, either as a national center or as a spiritual center. They have other axes to grind . . . .

It was always the same refrain at those teas, social gatherings, and dinners in British circles in Jerusalem, Cairo, and Baghdad, which I assiduously attended as a young correspondent and ex-British soldier, before I was tacitly dropped as an unreasonable Judeophile and an unorthodox because anti-imperialist, British subject: the Hebrew gold and the poor Arabs!

The poor Arabs with their national governments in Egypt and Iraq, their wealth in Syria—not to speak of the accumulated treasure of the Moslem religious foundations—could not match the wealth of the Hebrews! I recall one day, finding the Mufti of Jerusalem and his cousin Jemal Husseini in high glee: the Moslems of Java—forty-seven million of them—had just sent him a donation for the repair of the mosque of Omar, which had been damaged slightly in an earthquake.

"Figurez-vous," said the Mufti, "twenty-seven thousand pounds sterling have come in from the Dutch Indies, from Java alone, on my appeal. Last week a hundred thousand pounds from Afghanistan. And this is only the beginning: British India, Burma, Malaya, Singapore with its two hundred Moslem millionaires, Morocco, Algiers, and Persia are still to be heard from.

Wonderful isn't it? Now we, too, are going to build a university and schools and model farms like the Jews."

But the Arab university still exists only on paper . . . and the mosque was never repaired . . . and the Mufti is in Berlin, not badly off at all!

"In his capacity as watchdog," writes Humphrey Bowman, "the British official in Palestine, seeing Jewish immigration increase, began to share the Arabs' fear of Jewish domination."

Curious, isn't it? He, the British official, was there to further Jewish immigration, further the building of the Jewish national home, and he shares the Arab fear of Jewish domination. He calls himself a watchdog to see that the Jews don't get ahead too fast!

When the bogy of Hebrew gold made no impression on me (for I knew of the difficulties accompanying the collection of Jewish funds for Palestine from the Jewish poor in Poland and New York's East Side), another trick was used. Every one of the British officials in Palestine, Egypt, Trans-Jordan, and Iraq with whom I came in contact at one time or another whispered his suspicions that the system of co-operative agriculture practiced by the Jews in Palestine had Bolshevik roots. One had always to be protecting and explaining the Jews, not against or to Arabs, but against and to their official protectors, the gentlemen of the colonial service . . . .

The predominant sentiment in British administrative and military circles in Palestine and Egypt in the first years following the war was against the Balfour Declaration. The issuance of that document, the Magna Charta of the Jewish people, was deemed an unfortunate and regrettable wartime expedient. True, it had answered the purpose of its framers in that it had caused world Jewry to throw its weight into the scales on Britain's side. Therein lay its value. But what would be its consequences if its spirit was to be upheld in the postwar years? It was one thing to gain the support of the Jewish intellectuals and businessmen in time of war, but quite another matter to introduce, by virtue of the Declaration's implications, so heterogeneous and turbulent an element as the masses of eastern Europe into the but recently pacified Arab world.

The Jews had apparently taken the Declaration at its face value; they came streaming into Palestine from the four corners of the world. With the Arabs one could deal. They were not very different from other colonial peoples with whom one had experience. They were on the whole a submissive people. They had been taught their place and station in life for six hundred years by their Turkish masters. Their notables were polite and affable to the point of subserviency. They had picturesque manners. They were good sportsmen. They did not bother their heads about abstract questions of philosophy and economics, so long as the income from their feudal estates flowed in regularly. It was a pleasure to spend a week end in the home of an Arab prince, something to write home about. Britain wanted the Arabs as friends, didn't she? Well, they were her friends if Britain would just leave well enough alone. Why should she then want to play up to the whims of Mr. Balfour and a few other eminent statesmen who had no experience in dealing with Oriental peoples and to inject into the but recently tranquilized

Arab world these hordes of young Jews and Jewesses from Europe?

Did they think in London that this was going to improve British interests? The *status quo* had just been re-established with infinite pain and trouble. Englishmen occupied all the positions of authority formerly held by Turks. But those Jews, they were the eternal challengers of the *status quo*. They came with the definite intention to kill the *status quo* and make a new start. They had ideas about building a new life, a new society, a new world. They were eager, enthusiastic, zealous. They wanted to do things, change things, improve things. They would inevitably, if allowed to come in unchecked, acquire a voice in the running of affairs by the sheer volume of their numbers. They were bound to question and upset the old medieval property relations between the Arab landlords and their serfs. They would put notions of freedom, of a living wage, of popular representation in government into the heads of the fellahin. Not necessarily by agitation and propoganda, but simply by their example. Was that wise? Would that contribute to social peace in the Near East? Would that please the Arab nobility whose friendship Britain had but recently acquired and was eager to hold? Was it fair, too, to let France take over the administration of Syria, after British arms had conquered that country? Was it a statesmanlike procedure to allow Arab solidarity, brought about with so much difficulty and still precarious, to break up again by allowing alien and untried creeds of government to be introduced?

Such and suchlike considerations haunted the brains of the British gentlemen in Egypt and Palestine. They were fundamentally and *a priori* opposed or at least out of sympathy with the main project of British and international policy in the Near East: the establishment of the Jewish national home in Palestine. They were not anti-Semitic, although there may have been some vulgar Jew-baiters 'mongst them. They believed the establishment of the Jewish national home to be *au fond* detrimental to British imperial interests. They hoped that the Arabs would protest and if necessary rise against the French administration in Syria and throw the whole Near East into bloody chaos if need be, if only to convince the peace conference and the European statesmen that the provisional arrangements in Palestine and Syria were impracticable.

The Colonial Office functionaries could not very well come out openly with their objections and arguments. For it was their own government which had made the pledge of the Balfour Declaration to the Jewish people, and that pledge had international sanction. Britain was not to act in her own interests in Palestine, but as the chargé d'affaires as it were for the whole civilized world. The allotment of Syria to France had come about as the result of long-drawn-out negotiations with Britain's French ally. These things were settled in the regular, normal, diplomatic manner, with the approval of America. It was difficult and well-nigh impossible to upset the validity of these international agreements by egotistical, imperialist arguments.

Even the colonial functionaries in the Near East were aware that Great Britain could not lightly go back on pledges given freely and within earshot, so to speak, of the entire comity of civilized nations without acquiring an odious reputation for trickery and unreliability and without setting a dangerous precedent at a time when the wave of idealism set in motion by Woodrow Wilson was running high and the general expectation was that the world would be rebuilt on the foundation of the sanctity of pacts, sacred promises, and solemn new covenants. Even a Great Power could not, on the spur of the moment and under those circumstances, it was realized, repudiate the signature of its own Foreign Ministers at the bottom of important international documents on which the ink was scarcely dry.

There were still some moral inhibitions in that world of the early nineteen-twenties which checked the utterly cynical application of power politics which Hitler was to teach the democracies. The technique of making pacts under oath with those it was intended to destroy and emasculate through candor was not yet invented. The Great Powers still maneuvered with caution and behind a mask of benevolence. Woodrow Wilson and Masaryk, Balfour and Clemenceau were still alive. The governments tried not to soil their own hands with injustice, but left the more sordid detail to underlings. For Machiavelli after all, had taught long ago that there never is wanting a legitimate reason for the nonobservance of the plighted word to a prince who knows how to substitute for the strength of the lion the craftiness of the fox.

If those agreements in Near Eastern affairs appeared unassailable because they were backed by international sanction, they could still perhaps be shaken and ultimately even voided if approached indirectly and in a roundabout manner. For instance, if the Arabs could be made to see the reorganization of Syria by France and the settlement of Jews in the Holy Land as a breach of faith, as a great injustice imposed upon their race, an entirely new light might be thrown on the matter and entirely new prospects be opened. If they could be roused from their political lethargy and be made to protest and agitate and rebel against the provisions of the Balfour Declaration, and thus plunge Palestine and the Near East into a state of unrest and turmoil, would the world not come to see the folly of the scheme of those philo-Semitic idealists as Balfour, Lloyd George, Clemenceau, Smuts, Wilson, and Masaryk? Would that not irrefutably prove the unworkability, the impracticability, and the impossibility of the Jewish national home? Wouldn't the whole world see that the insertion of an island of Jewish activity in the Arab world was an error, a political mistake that could be rectified only by a revision of British policy, by qualifications of the international sanction? If there were enough unrest, strife, riots, and bloodshed, wouldn't the world step back in revulsion and recognize the cancelation of the pledge contained in the Balfour Declaration as opportune, justifiable, and wise?

That, in the main, has been the devious and unhandsome policy pursued by successive British administrations in Palestine and in the Near East for the past twenty-five years. Like a worm in the bud it has gnawed its way into Arab-Jewish relations until today its ravages must fill the most hopeful with dismay and it has become clear that only the prompt application of the most radical measures will be able to check and eradicate the blight.

The policy of corrupting Arab-Jewish relations was never avowed, always denied, yet it went on uninterruptedly and in unvaried form for years. Critical or remonstrative references to it were contemptuously dismissed by British statesmen and their partisans as being incompatible with and unworthy of a gentleman's ideal of carrying on public affairs and therefore impossible. And the Palestine administration's was indeed so fantastic a performance that it seemed unbelievable even to Zionists, who, in spite of evidence piling upon evidence in the course of the years that it was Britain's own agents who consistently sought to poison Arab-Jewish relations at the root, preferred to look elsewhere for the causes of Arab animosity. The great leader of the Jewish national cause, Dr. Chaim Weizmann, does not even today believe that Great Britain has betrayed him and the Jewish people.

And who indeed would dare accuse England, the sponsor of the Balfour Declaration, the generous mother of the free, who had taken upon herself the burden of Palestine, who was, moreover, the traditional protector of the Jews, of such duplicity, of acquitting herself of the task entrusted to her by the civilized world in a manner so dishonorably at variance with the legal contract? To what purpose would England betray the Jews? What could she gain from it? Wasn't it Britain herself in the final analysis who paid for the general unrest and the periodic outbreaks of violence in blood and tears and treasure? Could such things really be?

For a quarter of a century Britain has played the role of the magnanimous big brother whose sole aim it was to bring peace and prosperity to the poor benighted Arab and, at the same time, to be a gracious host to the Jews, but whose purely altruistic designs were always frustrated by that commitment to allow the Jews to build their national home and ultimately to have an independent Jewish state in Palestine. All the big brother's good will and all his impartial benevolence broke and was spilled on the rock of Arab opposition to that enterprise. No matter how hard he tried, by cajolery, by force, by argument, or by temporizing, he never succeeded in overcoming that fundamental Arab intransigence, the obvious fact of Judeo-Arab incompatibility, or, *quod erat demonstrandum:* the temperamental unfitness of the two to live and work together.

Jews and Arabs were nowhere and never in trouble since the days of Mohammed. The Prophet issued special injunctions to his followers to deal honorably, justly, and kindly with the Jews, because like the Arabs they too were the children of Abraham, the witnesses of the past and the People of the Book. History, in fact, shows few instances of a more peaceful and ennobling meeting of two nations on one spot and of the elevation of lingual and national differences to brotherhood by mutual respect and appreciation than that of Jews and Arabs in the Baghdad of Harun al-Rashid and later in Spain. From that brilliant and beneficent collaboration there issued one of the parent streams of Western civilization. Under the Turks, Jews and Arabs had always dwelled together in unity. Why then could it not be so under the supposedly far more liberal and humane rule of Britain?

It could not be so because Englishmen began to "preach nationality." It could not be so because from the very start Englishmen belonging to the government services were sowing the seeds of dissension in Palestine by teaching the Arabs the rudiments of chauvinist nationalism: suspicion and hatred of the stranger. The establishment of the Jewish national home, the Arabs were told, would obviously mean a great increase in Jewish immigration and their own economic and political subjugation in the end. Before a civil administration was set up, and General Bols still ruled in Jerusalem, and long before the mandate over Palestine was entrusted to Great Britain, the foundations of that moral antagonism which was to widen with the years until it appeared almost unbridgeable were being laid down by nationals of the Great Power which, in the Council of the League of Nations and before the Permanent Mandates Commission, was to cite a congenital Judeo-Arab incompatibility as the reason of and the excuse for its failure to have brought social and racial peace to the Holy Land.

Yet there is good reason to believe that the mass of the Palestinian Arabs, destitute and miserable as they were, instead of resenting the earlier phases of Jewish immigration, rather looked forward to a still greater influx of Jews with eager anticipation. The hope of sharing in the material benefits that would naturally accrue to all the inhabitants of the land from a large-scale settlement and the modernization of the country was still lively. For it appeared obvious that in order to succeed the Jews first had to clear the ground for their national home, that is to say: drain the swamps, reforest the soil-denuded hillsides, combat disease (especially malaria, tuberculosis, and trachoma) by establishing hospitals and medical centers, build roads, dig wells, lay out new plantations, and generally clean up, in all of which activities the Arabs were to share and to be paid for. When therefore Feisal, who in the most important negotiations with the Great Powers and at the peace conference had been the chief Arab spokesman, instead of raising his voice against Jewish immigration into Palestine, welcomed the Jews publicly to the Near East and hailed their return as the beginning of a new era of peaceful collaboration and well-being for all, not an Arab in Palestine, or anywhere else, contradicted him by as much as a word. The Arabs were not only sympathetic to the idea of seeing the Jews start to work in the land that was now officially rebaptized Eretz Israel, Land of Israel; Feisal said openly that they were eagerly awaiting them.

"We," on the other hand, that is British civil servants, wrote Charles B. Ashbee*, civil adviser to the Governor of Jerusalem, Sir Ronald Storrs, in a letter dated July 25, 1918, regarding his own and the attitude of his colleagues in the Palestine administration towards the projected Jewish national home, "are preaching nationality in Palestine . . . . We are for the Arabs . . . . We make great capital out of the Arab tradition that Jerusalem comes back to the Arabs when a new prophet shall enter as a conqueror." Who is is this new prophet?

*Quoted by Paul L. Hanna, in *British Policy in Palestine,* American Council of Public Affairs, Washington, D. C., p. 173.

Mr. Ashbee took advantage here of a play of words and of the general illiteracy of the Arabs, to whom the words Allenby and al-Nebi (prophet) sounded quite similar when rapidly pronounced.

When men like Ashbee spoke of Jerusalem coming back to the Arabs they not only betrayed their antipathy to the Jewish settlement of Palestine which they themselves were to supervise and facilitate, they also had their eyes on Syria, where the French were installing themselves. That reference to Jerusalem was a sly appeal or hint to the Syrian Arabs. The British functionaries still hoped for the Arabs' sake, it goes without saying, that instead of French control of Upper Syria and a Jewish settlement in Lower Syria (Palestine), there might emerge from the unsettled political conditions a united Syria in which neither the French nor the Jews would have any business, but only the Arabs and of course their good friends and guides—the British. Some (British) officials cannot have failed to hope, writes one observer** of the postwar phase of Near Eastern politics, that events, if left to themselves, would present the world with a *fait accompli* in which both France and the Zionists would disappear from a Near East where British dominance would be assured.

When a Zionist Commission, made up of prominent British, French, Italian, Dutch, and Russian Jews and headed by Dr. Chaim Weizmann, the eminent chemist who had just rendered Britain an important service, was granted permission to visit Palestine and arrived in that country in April, 1918, its members were at once disagreeably surprised to find not Arab hostility,

**Ibid.

but themselves viewed as interlopers and a nuisance by the British authorities. They were snubbed and given the run-around, as we say nowadays, as if they were costermongers who had invaded His Majesty's private garden party. The Zionists soon discovered that the approbation and the enthusiasm of great Britons like Lloyd George, Balfour, P. G. Scott, and Borden and of non-Britishers like Clemenceau, Tardieu, Smuts, Masaryk, and Orlando for the redemption of the Holy Land by the Jews was by no means shared by the clique of military and civil administrators from Egypt. Where they had expected assistance, they were met with rebuffs and downright obstruction. The Zionists tried to have Jews employed in the public services, strove for the recognition of Hebrew as one of the official languages, and made a few more demands of the kind legally warranted to expect prompt fulfillment. They merely wanted to start the ball rolling, so to speak, after having been given the starting signal in London.

The Palestine administration did not tell the members of the Zionist commission that it considered them intruders and unwelcome busybodies who were snooping around where they had no business. It reproached the Zionists with impatience. Rome was not built in a day, these men were to remember, and a great undertaking of unprecedented specifications like the building of a Jewish national home would first of all require careful and painstaking planning and preparations. The Jewish people had waited two thousand years for Palestine? Could the Zionists not wait a few months longer? Why all this unseemly haste? One had to proceed step by step and perhaps increase the tempo of building as the work got successfully under way.

And then, of course, there were the Arabs. The Zionists had to remember that the Palestine administration was to guard their sensibilities, too. The Arabs must not gain the impression that they were going to be swamped by a flood of Jewish immigrants. If London looked with favor on the establishment of a Jewish national home in Palestine, it was to be remembered that Britain must also protect the civic and religious rights of the indigenous population. The Zionists might just as well make up their minds at once that the Palestine administration was not going to be party to the least injustice inflicted on the Arabs. To all of which the Zionists might have replied—and probably did—that they had no intention of infringing upon Arab rights whatever, that these rights were as sacred to Jews as they were to Englishmen, that no Jew expected or wanted to obtain justice for himself and his people at the expense of others, and that the only things they expected to take away from the Arabs were their diseases and their poverty.

They might have said these things if they had seen that there was any use in saying them, if they had not soon become aware that the lot of the Arabs was not the issue involved at all, and that the Palestine administration was talking one way and thinking something entirely different at the same time. The Palestine administration was engaged in playing the classical diplomatic game of gaining time.

By using that tactic of passive obstructionism the administration aimed at keeping the situation in Palestine fluid, undecided, and up in the air, so to speak, and everybody in suspense and on tenterhooks. Nobody, Arab or Jew, was to know what exactly was going to happen or what to expect. Not that these gentlemen in the administration were themselves at sea or dubious of the future, although they pretended to be so. They knew full well what Britain was expected to do in the Holy Land and what the home government intended to do. They were well aware of the fact that support of and even enthusiasm for the building of the Jewish national home was growing by leaps and bounds, especially in Nonconformist and Free Church circles in England. They knew that in the international diplomatic congresses Lloyd George used his votes in favor of the Jewish homeland by the frank declaration that the Methodist churches of Wales favored its establishment.

The idea of England sponsoring a project that would fulfill ancient prophecies appealed deeply to the imagination of Protestant fundamentalism everywhere. There were men like Smuts of South Africa, Colijn of Holland, and Masaryk of Czechoslovakia, all three in the Calvinist and Hussite tradition, who did not hesitate to say that if nothing else came out of the Great War but an end to the national homelessness of Israel, the struggle would not have been in vain. The men in the Palestine administration felt this world-wide moral pressure. They knew that it would inevitably be translated into action unless a strong opposition developed. That is why they temporized and postponed and dallied with the application of elementary measures of foundation laying. That is why they acted churlishly towards the Zionist commissioners and reproached them with undue haste and impatience and forwardness. They themselves did not want to settle anything. They wanted to hold up the show as long as possible, so that dissatisfaction and tension might steadily develop and finally erupt in an explosion that would rock Palestine from one end to the other. They wanted to see the flames leap over into Syria. When that happened, the home government would certainly pause in its headlong drive to put the Jews into the saddle in the Near East and reconsider Balfour's utopian scheme and maybe call off the Zionists and all those Jewish zealots and enthusiasts who were now beginning to arrive with every boat from Eastern Europe.

Would that the Palestine administration in those early days had acted in good faith, in the spirit of Balfour and the British government of which he was a member, and the Jews had been told frankly and generously. here you are! This is your national home! The doors of Palestine are open! Take the available land, the vast areas of the former Turkish crown lands, on which not a soul has been living these hundreds of years and which have turned into wilderness by neglect. Take that land and cultivate it! Make it productive! Plow up the weeds and cacti! Lay out your orchards and wheat fields and orange groves! Make them yield a hundredfold by modern methods of agriculture! Build your cities on it: a new Jerusalem, a new Hebron, and a new Jericho. Erect your schools! Let your children play in the new parks and forests of eucalypti. Tell the Jews of Poland and Russia and Rumania to leave their stinking ghettoes at last and come over to breathe the free air of their own land. Tell them that their prayers have been answered! Tell them that their ancient longing will be stilled, because England wills it so. Let there come as many as the land will hold until Eretz Israel is filled again as of old, from Dan to Bersheeba, with the children of Israel. Let the weary and persecuted forget the memory of the evil years and the centuries spent in the blood-bespattered Dispersion. Let them be at home in their own national home! Here a new world and a new life is open to them! Only see to it that no ill befalls these starvelings of Arabs, the fellahin peasants, who are now the slaves of their landlords and moneylenders! Let them share your new freedom!

If the Palestine administration had spoken that way, in that spirit, the spirit of the Balfour Declaration, as it should have spoken and as the civilized world expected it to speak, I make bold to believe that we would never have heard of riots in Palestine, of bloodshed and terror, and of those silly, pompous, fact-finding commissions which did not want to find facts leading to the truth, but which merely covered up with whitewash the men in the Palestine administration primarily responsible, consciously and willfully, of all the misery and pain and heartbreak that have accompanied and gone into the building of the Jewish national home in Palestine these past twenty-five years.

By the end of 1919 disturbances had broken out in Palestine, that is to say before the peace conference, which was finally to settle the status of the country, had begun its sessions. Arabs were killing Jews. Jewish property was being destroyed and looted under the eyes of the Palestine police, which, it has been formally established, refused to interfere. At the same time, the situation had grown so tense in Syria that France decided to reinforce her army with some twenty-five thousand fresh troops. The same British agents, members of the colonial service in Egypt and Palestine, who had brought the situation to a head in the Holy Land, were egging on their friend and protege Feisal (the third son of King Hussein of the Hejaz), who had served with Lawrence in the liberation of Damascus and who had been placed in charge of the provisional government in that city, to make a determined effort to exclude the French from Syria. If he proclaimed himself King of Syria and Palestine before the new French army arrived, and assumed the royal power by the backing of the Syrian nationalists, both Great Britain and France would face a *fait accompli,* which could not be undone except by force.

Counting on the British government's reluctance to employ violence and its anxiety not to quarrel with France, the Zionist experiment would then in all likelihood be liquidated or greatly curtailed and France would be persuaded to restrict her interests in Syria to cultural and commercial representation: that is, leave Feisal in possession of the throne with, at the worst, some indirect French control through a high commissioner or an advisory council. The unfortunate Feisal fell into the trap and assumed the title of King of Syria and Palestine. There was great rejoicing in Damascus and in the mosque of Jerusalem, which latter city was at last to come back to the Arabs.

None of the parties concerned, however, had taken

into consideration the moods of the Tiger of France and . . . the Mosul oil fields. When I was Clemenceau's guest in later years and discussed Near Eastern affairs, he said, harking back to Feisal's royal proclamation: "I was perfectly well aware that Feisal was a puppet of *ces messieurs du Caire et de Jérusalem,* those gentle men in Cairo and Jerusalem and of their plot to throw us out of Syria and rob the Jews of their birthright at the same time. So I took the decision *de foutre l'Arabe à la porte,* to throw the Arab out (of Damascus) at the first opportunity. In an acrimonious exchange of compliments with Lloyd George, my colleague, I forced the British government to notify Feisal that changes in the status of the Near Eastern countries could only be made by the peace conference. When the King thereupon still hesitated, I ordered the army into Damascus and chased the fool back to his old father, that still greater fool in Mecca. All that *ces messieurs* could do for him after that was to send a guard of honor to salute him at the railway station of Ludd when he passed through on his way to his senile pappy."

Clemenceau had merely insisted that France share in the oil production of Mesopotamia and control the Syrian territories over and through which the oil was to be piped to the Mediterranean . . . .

The British were soon to find another throne for Feisal in Baghdad. In the meantime the Syrian nationalist agitation soon petered out. France made so formidable a show of force that the Syrian revolutionary committees saw the futility of prolonging resistance. General Gouraud's military preparations demonstrated two things: first, that France was in dead earnest about acquiring Syria as her sphere of interest in the Near East plus a share in the Mesopotamian oil output, and secondly, that the Syrian nationalists had been lured into an extremely dangerous game when they listened to suggestions emanating from British quarters in Egypt and Palestine that France could be easily eliminated if only they (the Syrians) resisted her with determination. The Syrians were left holding the bag after the gory adventure. Any hope that the British government would come to their assistance by diplomatic support of their claims for independence was dashed to the ground when Lloyd George told Feisal to bow before the French demand to vacate the throne of Syria.

The first act in the duel of empires for predominance in the Near East, which was to last until 1941, when Britain, as the result of totally unforeseen circumstances occupied Syria by defeating the forces of Vichy, ended with her as the temporary loser.

The second phase in that interimperial struggle opened in 1924, four years after the dismissal of Feisal, when the Djebel Druse tribes rose against the French mandatory power. Their revolt spread so fast that it won over the popular masses in Damascus and Homs in a few weeks' time with a result that the French, who, if not taken by surprise, were at least thrown off their balance by the magnitude of the uprising, were in danger of a debacle. Troops were summoned from Algiers and Morocco, but before they could arrive a number of French garrisons in the interior of the country had been massacred and several relief columns sent out to restore order had been waylaid and put to the sword. For a year the country was in convulsion.

I happened to be in Damascus when the situation was at its worst as the guest of my landlord, the owner of the house my family occupied in Bourg-en-Forêt in France, M. le vicomte Charles Henri, colonel in the sixteenth regiment of dragoons, stationed at Saint-Germain-en-Laye, who was temporarily attached to the intelligence service in Syria. At the Grand Serail, the former palace of some pasha, now general headquarters, Colonel Henri told me on October 4, 1925, that the presence of French civilians and all Occidentals could no longer be countenanced in the capital city. There were rumors of *une grande nuit de couteaux,* a big night of knives, being plotted in the Choggur and Meidan quarters. As memories were still very vivid of the general massacre of Christians carried out in 1886, the French High Command itself had become nervous.

"But this has been going on for a year," I said. "Surely, there will be sufficient reinforcements in hand soon to take the situation in hand. Paris is surely not going to wait till we are all murdered here in a new kind of Sicilian vespers?"

"Mon ami," said the Colonel, "you are as naive as the members of the French Government. M. Briand (the Foreign Minister) will not believe us, will not believe General Sarrail (High Commissioner in Syria) that the situation is serious."

"You mean the French army cannot handle this handful of Druses?" I ventured.

"It isn't that," he replied. "In the first place, there isn't any French army. We have never received the reinforcements we asked for: all reinforcements are being diverted to Morocco (where Abd-el-Krim of the Riff tribes had risen in revolt). Secondly: we do not face the Druses alone. We know that twenty thousand rifles and ammunition have just been smuggled into the country from Trans-Jordan and Palestine."

"By the British?" I asked incredulously.

"By the British!" assented Colonel Henri.

"But M. Doumergue (President of the French Republic) and King George (of England) are exchanging—or are about to exchange—visits to reaffirm the ties of the old Entente Cordiale. Britain and France were never such good friends as at present . . . ."

"Officially, yes," he smiled. "But here: *on nous poignarde dans le dos,* we are being stabbed in the back . . . . Every Druse woman is wearing a new necklace of British gold coins around her neck. The English have mobilized their guineas and sovereigns, 'the Cavalry of St. George,' as we call their money here . . . . Briand has spoken to the British ambassador in Paris about this business. But that gentleman furnished our Foreign Minister with verbal proofs that all is well. In the meantime we here are exposed to the tortuous machinations of a few zealous agents of the Colonial Office and a few British officers."

"Then what do they want?" I asked.

"Want? *Eh bien,* this Druse revolt is to spread to all Syria and then invade Palestine. We are to be shown up before the world as incapable of administering the mandate. We are to get out, that's all. Then the British will restore order. Syria is to be united with Palestine for the Arabs, *c'est à dire* for the British! That's the whole scheme in a nutshell."

I told Colonel Henri I wouldn't leave. "If necessary," I suggested, "you can give me a French uniform to wear. I'll stay with you. France is in the right here, and so are the Zionists in Palestine. You cannot allow yourselves to be stampeded."

He laughed. "Have it your own way," he said. "You are a British subject and may be useful if only to go to Palestine and tell *messieurs les Anglais* that we know their game . . . ."

I watched the bombardment of the Meidan and Choggur *souks* and the attack on the Azem Palace on October 19 and again on the twentieth. Fires had been started everywhere by the rioters and looters. The whole city of Damascus, from Fort Gouraud to the *souk* of Midhat Pasha, bristled with entrenchments and barricades. Armored cars ventured in at times if only to reassure the Christian population. But no sooner had they proceeded a few blocks when rifle fire broke out from all the terraces and windows. It became clear that it would be impossible to quell the disturbances by street fighting, which would have necessitated the capture of the revolted quarters one by one. Apart from the fact that this method would have required the employment of very strong forces, which General Sarrail did not have at his disposal, it entailed risking the lives of the few troops available and, of course, would result in great material damage.

For three days and three nights the ammunition used in the barrage intermittently thrown between the Azem Palace and the quarters which were the heart of the insurrection consisted solely of star shells. Not until the night of October 19-20, when the insurrectionists began to launch mass attacks which bore all the earmarks of being directed by European strategists, did General Sarrail order genuine shrapnel poured into the Meidan. This bombardment, which had a ghastly effect (for it was a rain of fire into the most crowded bazaar section), lasted until dawn. Shortly after nine the Syrians asked for a parley and at noon surrendered their firearms. But that did not end the rebellion. It merely meant that Damascus had been pacified and that the Druses and Transjordanians, who had for months been filtering into the great city, withdrew to the hills. In 1927, when I visited Syria again, France still had seventy-five thousand soldiers in the country, and raids and desultory fighting was still going on.

Immediately after the bombardment of Damascus in October, 1925, an outcry was raised in the reactionary press in France against "the massacre of the noble Druses" by the "Freemason Sarrail." The British government complained that three British "protected persons" had been killed in the disturbances . . . and General Sarrail was recalled.

I had known these three British "protected persons" while in Damascus. I met them at the British consulate where I had to go and call frequently myself in order to obtain visas to go in and out of Syria. The curious thing about their death was that they were shot from the trees in the square of the Grand Seraglio, which was used by the French as general headquarters. The three British "protected persons," all three of them Arabs from Trans-Jordan, were among the snipers who began the outbreak on October 17.

In the spring of 1927 I visited the retired General Sarrail in his home in Montauban and asked him about the origins of the Syrian affair. "Some of the Parisian newspapers," I said, "accuse Russia of having fomented the trouble."

"I have read that, too," he replied. "But I know better. I know as far back as 1924 that the revolt was coming. I had been forewarned. No less a person than Abdullah, the Emir of Trans-Jordan, a British puppet, volunteered the information that officials of the Colonial Office and certain British officers in Palestine were trying to make trouble for us in Syria."

"Why should Emir Abdullah," I asked, "who had but recently been elevated to the throne by the British government in a country that was arbitrarily detached from Palestine, wish to betray his benefactors and protectors?"

"Oh, that was a question of inter-Arab intrigue," came back the General. "Abdullah was angry that he was not employed by the British, and that they had selected, instead, a rival of his, an influential Trans-Jordan tribal chief . . . . Abdullah, moreover, wanted to get into our good graces, figuring that we might restore the monarchy in Syria, from which we had ousted his brother Feisal four years before, and give it to him. We had not the slightest intention of doing this, of course. Abdullah acted on the presumption that when one goes fishing in troubled waters one never knows. He did not like his fly-bitten little capital city of Amman very much and was looking for something better. He played with us in the hope of Damascus. He played with the British in the hope of Jerusalem."

"Jerusalem?" I asked.

"Evidently," came back Sarrail, "that is the significance of making him Emir of Trans-Jordan. Some day the Arabs in Trans-Jordan and Palestine will clamor for such a reunion, or be made to clamor for it, which amounts to the same thing. The two countries, now separated arbitrarily, do in fact belong together. They are both Palestine. It is merely that the Jordan divides the country into two parts. When the British Colonial Office thinks they are ripe for reunion, it will set the Arabs yearning for reunion. Their separation will be made to appear unbearable."

"When will they be ripe for reunion in that way, do you think?" I asked.

"Well, just as soon as the British think they need Trans-Jordan and its Emir to constitute a counterbalance to Jewish influence in Palestine," said Sarrail.

"But Palestine is to be transformed into a Jewish homeland by decree, of the League of Nations. France ceded her interests in the Holy Land on the specific understanding that the pledge to the Jewish people was to be carried out."

"Monsieur," broke in the General, "what can the Jews say if and when the Arabs of the two countries want to reunite? What argument can they have if England demonstrates to the whole world that the Arab claims are legitimate and that if they don't give in to these aspirations there will be trouble?"

"The Jews could lodge a complaint with the Permanent Mandates Commission of the League of Nations and say that the Arabs are put up to this by British agents," I replied.

"Non, monsieur, they cannot," shot back General Sarrail. "Any complaint the Jews want to lodge against British intrigue must pass through British hands first. The Jews have no right to appeal to the League or to any agency of the League. That is the mandatory power's prerogative. Can you see Britain launching a protest in Geneva against . . . Britain? *Ma foi,* I can't!"

"The world can be told about it," I said.

"Doit-on le faire? Should it be done?" asked General Sarrail. "That is the question I am asking myself all the time. Should we contribute to a deterioration of Anglo-French relations by making public the perfidious role played by Britain in the Near East against France and against the Jews? Would we serve the cause of peace? I myself," he continued, "have absolute proof here in my house—and there is absolute proof in the second bureau (of the intelligence service) in Paris that British officers fomented the troubles in Syria, supplied the rebels with arms, and even directed the insurrection for the purpose of expelling France from Syria—and I don't know whether to use the information or not."

"How many lives did it cost France, the Syrian insurrection, I mean?" I asked.

"It cost us tens of thousands of men, scores of thousands . . . . " he said.

"Has Your Excellency nothing to say for these Frenchmen? Did they know?"

"If I were to publish the evidence I still have here in my files," General Sarrail went on, pointing to a steel cabinet. "I would most probably precipitate a fierce diplomatic incident the consequences of which might be quite serious . . . . I am subjected to fierce attacks in the clerical and reactionary press in Paris for what is called my mismanagement of affairs in Syria, for the bombardment of Damascus, for having insulted, as they say, in my quality as a Free Mason, the noble Christian population of Lebanon, for having spread republican ideas and I don't know what else . . . . I could silence all that slanderous drivel with one word or with one gesture. I could lay before the public my evidence and the evidence available in the Second Bureau and prove that British officers fomented the trouble in Syria, that it was they who supplied the rebels with arms and even they who directed the uprising . . . . "

"For what purpose could they have done that?" I asked. "Were those British officers perhaps freebooters, men who were imbued with ideals of freedom and who wanted to see Syria gain her independence? There have been such cases . . . . Some Britons fought on the side of the Boers against Methuen and Kitchener in the South African war . . . . "

"I have never heard of that," came back General Sarrail dryly. "I can assure you that in Syria we were not faced by soldiers of fortune. They were members of the British Palestine administration, of the British military forces stationed in Trans-Jordan, and behind them stood the British agency in Cairo. And their purpose and intent was to show the world that France is incapable of assuring security in Syria, incapable of exercising the mandate over that country. They wanted to prove that she should get out so that Britain can wield all authority in dealing with the Arabs . . . . "

"But in Palestine," I ventured to object, "the recent troubles in Palestine between Jews and Arabs were also fomented by the British. At least so I have heard in different quarters. The British surely are not working to have themselves expelled . . . ."

"Of course not," replied the General. "They want the Jews to get out . . . ."

"But Britain brought the Jews in herself," I objected again.

"The British also brought the French into Syria. The British conquered Damascus, do not forget. In the international conferences Britain agreed to cede Syria to France and Palestine to the Jewish people. Both the French and the Jews were Britain's allies in the Great War. They were entitled to a share in the distribution of former Turkish territories. Britain acted as if she had acquiesced in the reorganization of the Arab countries. But in reality she thought: *très bien,* you go in there, you Frenchmen and you Jews, we must, alas, recognize your right to do so, but we will soon make it impossible for you, and you will be only too glad to get out and then we will be rid of rivals. *C'est très simple.* It just takes a few good secret agents to cloud the issue as we saw in Syria . . . ."

"I still cannot imagine, *mon général,*" I said, "why you do not vindicate yourself in France."

"At this time when the international situation is again growing dark?" the General took me up. "What would I gain?: Vindication for myself? But serious complications might follow for France, which stands alone on the Continent *vis-à-vis* a Germany thirsty for revenge. No, I shall have to wait. The truth will come out some day. The last word has not been spoken yet in the Near East. Britain never gives up. She will return to the task. England was my implacable enemy, but I can hold my head high, for if she pursued me with her hatred I cannot think that I am a bad Frenchman . . . ."

General Maurice Gamelin, upon his return from Syria, where he had succeeded Sarrail as military commander (though not as High Commissioner), addressed to the Ministry of War a letter and a report containing some observations on the Druse revolt and its causes. The letter and the report, drawn up at the headquarters of the military intelligence service in Paris, were dated July 20, 1926. They were not made public until the year 1929, when the Near East again started to flame, with Palestine as the hearth of the conflagration. Sarrail's private secretary, Paul Coblentz, took it upon himself to release the documents.

Here is the letter from General Gamelin:

July 20, 1926

To the High Command of the Army of the Levant
General Staff, Second Bureau
SECRET
General Gamelin, Commanding the Army of the Levant to the Minister for War:

I have the honor to transmit to you herewith a copy of statements made by Chief X, who surrendered unconditionally.

X has from the beginning played a leading role in the rebellion. He has certainly been kept informed of all the political intrigues which have accompanied it.

His statements appear to be sincere on most

points, and especially on all those which concern the activities of the other chiefs in the rebellion.

They throw fresh light upon the beginnings of the rebellion, or rather they confirm the indications that have already come to our notice, particularly with regard to the systematic campaign of agitation carried on by the People's Party with which Soltan (the Druse leader) has allied himself since the spring of 1924, with a view to producing the incident that would give the signal for the rebellion; and upon the activities of the Trans-Jordanian authorities, and perhaps of certain British officers. At the time we found ourselves facing grave dangers in Morocco, and it would be interesting to know whether the same influences have not been at work in both cases.

I have the honor to request you to be good enough to forward the enclosed copy of this interrogatory to General Sarrail, since the quoted statements of the Druse chief are concerned with the period of his command and bring out clearly the real origin of the rebellion, the causes of which certain adverse criticism has claimed to find in what were in reality mere pretexts.

X, moreover, constantly refers to activities favorable to the rebels on the part of the authorities of Trans-Jordania and Palestine, both native and British. Much information, almost certainly reliable, had already come to our notice with regard to the same facts.

There is no doubt that the neutrality of our southern neighbors has been entirely benevolent towards the rebels. In Trans-Jordan and Palestine appointments in the native administration are held by partisans of Feisal and by former Turkish officials and officers, in great part hostile to France.

The British officers and officials belong to that colonial administration whose spirit is well known. They are quite capable of paying no attention to instruction given by highly placed authorities, so long as they believe that in so acting they are serving the interests of their country. The only means of constraining them to drop their line of action is to collect adequate proofs and place these before their chiefs.

(signed) GAMELIN

The Druse chief interrogated by the French intelligence service was a member of the Djebel Druse Representative Council and belonged to a very old Djebel family. Intelligent, and little occupied with politics, he was reckoned among France's best friends. He was forced by Soltan to join in the revolt at the time of General Michaud's march on Soueida.* Being very brave and an excellent horseman, he immediately became one of the rebel chief's principal lieutenants.

He played a leading part in all the important incidents in the Djebel Druse region, then surrendered, and was banished. His cross-examination, an official document of the Directorate of Native Affairs in the Levant, was forwarded by General Gamelin to the Ministry of War with the letter which has just been quoted.

*One of the French generals whose column was attacked and destroyed when the Druse suddenly revolted in 1924.

Here are some specially significant extracts from this document:

Question: How and by whom was the attack on General Michaud's column organized?

Answer: Before the battle of Mezraa, and while he was still hesitating to join the movement, one of the principal chiefs of the People's Party at Damascus asked Soltan how he could make war on the French, seeing that the Druses had neither munitions, arms, nor equipment. It was then that Chahbandar said he would undertake to provide munitions and all the war material, and indeed he left for the south and made his way to Fedein in Trans-Jordan. Seven or eight days after this meeting he returned, bringing supplies of flour and food. It was the English themselves who sent these supplies by rail. The English carried the food and munitions by rail to Fedein and they were then brought by camel to the Djebel.

Question: Were there not foreign officers among Soltan's staff?

Answer: Yes, there were foreign officers. I often saw them with Soltan Attrach. I asked where these officers came from. I was told they came from Emir Abdullah's army (the British army in Trans-Jordan).

Question: These were the men who organized the attacks against the French?

Answer: Yes. There was formerly also Fcuad Selim, who was one of the great chiefs and drew up plans.

Question: How did the Druses renew supplies

of food and munitions? Where did the munitions come from?

ANSWER: From Trans-Jordania, and arms are still coming from Trans-Jordania up to the present moment.

QUESTION: Give further details on this point. Who sells arms in Trans-Jordania and who leads the convoys bringing them?

ANSWER: The convoys of arms, munitions, and money arrive at Fedein by train. They are generally convoyed by a Trans-Jordanian officer. The convoys come from Amman. When they are unloaded at Fedein there is an English railroad employee who takes delivery and deposits them near the station. A Trans-Jordanian officer named Hassan Effendi delivers them to the Druses. Oglei Kitami, a Christian, and his son Moussa bring these convoys back to the Djebel on pack camels.

QUESTION: What route do these camel convoys follow? Which way do they pass and where do they arrive?

ANSWER: They used to come by the desert as far as Dibine, but now the convoys no longer come to Dibine, but to Azrak instead. At the railway station, the British guard sends armoured cars to escort the Druse convoys for a certain number of kilometers, to protect them against Bedouin raids.

QUESTION: Is this distribution made regularly?

ANSWER: Yes, regularly.

QUESTION: What measures should be taken to stop this importation of supplies?

ANSWER: It depends solely on the English. If

they were approached, and consented . . . .

QUESTION: You say, then, that if the English wished to stop it, they could?

ANSWER: Obviously.

QUESTION: What do you know of the relations between the Druses and the English? Have any English been in the Djebel?

ANSWER: After the attack at Kafer, an English officer of high rank came in a car from Trans-Jordan to Aere. He had a secret interview with Abdul Chaffar Pasha, Soltan, Fedallah Pasha, Huneidi, and Ayel Amer (rebel chieftains). On leaving Aere, this officer went to Kafer with Soltan and photographed the bodies of the killed. Soltan said he was a journalist, but the interpreter told me that he was certainly acting as an officer. Some days after this officer left Djebel, several cars loaded with munitions of war arrived at Aere. Soltan, who was there, declared that he was very pleased to see that the English officer had kept his promise . . . . From time to time, both before and after the fight at Mezraa, Soltan assembled the Druse chiefs, and read them letters he had received from Emir Abdullah, Rida Pasha, Rikabi, and an English officer from Trans-Jordan.

QUESTION: What did they say in these letters?

ANSWER: They encouraged the Druses and urged them to continue the struggle against the French. These people promised Soltan supplies of money, munitions, and reinforcements. The money arrived regularly. The reinforcements of men came only once, about August 20, and on August 24 they took part in the march on Damascus.

Since then no further fighting men arrived, only Sherifian officers. English officers came from time to time to see Soltan, who said one day that the English, being allies of the French, could not openly support the hostilities by permitting the despatch of reinforcements, but that they would continue to send him munitions and money. Camel envoys were bringing back from Trans-Jordan food, weapons and munitions.

QUESTION: What is the situation of the Druses who are now refugees in Trans-Jordan and Palestine? How many of these are there?

ANSWER: There are nearly five hundred Druse people now in Amman.

QUESTION: Are they armed?

ANSWER: The English tell them that if they want to return home to continue the war, their arms will be returned to them, but that otherwise they will be disarmed.

QUESTION: Do the English officers in Trans-Jordania make difficulties when the Druses pass to and fro across the frontier, whether armed or not?

ANSWER: At the frontier the English give the Druses their choice of returning to Djebel with their arms or of settling in the English zone unarmed. Those who remain in the English zone are well treated by the English, who give each man five Egyptian piasters a day.

Scarcely had Syria come to rest when trouble broke

out again in Palestine. In the month of August, 1929, on a preconcerted signal given in the mosques, and with the cry *ed-dowle ma'ana,* the government is with us, bands of Arab idlers and coffeehouse loafers, reinforced by the poorest of the peasant class and Bedouin, threw themselves on Jewish settlements, on Jewish property, and on Jewish individuals in various parts of the country. The outbreak was directed by the Mufti of Jerusalem who told me that the Arabs had their most sacred prepossessions violated by the Balfour Declaration and that they would go on attacking Jews, no matter how long it took, until Britain repudiated her policy of trying to establish a national home for the Jewish people in a predominantly Arab country.

However, instead of molesting the newcomers, the Zionist pioneers who had come to Palestine under the provisions of the Balfour Declaration and since Britain's assumption of the mandate, the Mufti directed his fury against peaceful Jewish communities in towns like Hebron, Safed, Tiberias, and the Bab Alchota quarter of Jerusalem—communities which had always existed in the Holy Land under Latins and Byzantines, as far back as the Emperors Titus and Hadrian, and under the Turks. He sought out those communities because his followers were bound to meet with little resistance from those pious old Jews, whereas in the newer settlements the marauders usually got a reception that did not tempt them to return.

Even so, the Holy Land was in turmoil. Life came to a standstill as the ordinary safeguards of government ceased to function. The mob, in itself not very numerous or impressive, dominated the scene. Sir John Chancellor, the High Commissioner, was on leave of absence, and his place at the top had been taken by Harry Luke, the Chief Secretary, as Acting High Commissioner.

This gentleman appeared to be in a blue funk in the first days of the rioting, when I called on him at Government House in Jerusalem. "The situation has gotten out of hand," he repeated over and over again, "and just at a moment when there are no troops within easy call."

I asked him if there were no arms on hand either. "Plenty," he said. "Well, then there is nothing to be afraid of," I suggested. "Give me a gun and give a few more British subjects a gun, and we can keep order. If need be, you could arm a few thousand Jews. They are being attacked by people who seem to have plenty of arms. It would be logical, in view of the fact that the government does not possess the means to protect them, that they be allowed to defend themselves."

"But that would mean civil war," objected Mr. Luke.

"Well, you pretty well have a civil war on your hands as it is."

"We, the government, would merely become a third party of disorder if I carried out your suggestion," he said.

"Disorder? By suppressing disorder you do not become disorderly," I remarked. "The Mufti's clique is the party of disorder. The Jews, whatever your objections to them may be, are the party of order. I have looked at the Arab rioters in Jerusalem, Hebron, Safed, and elsewhere. They are the lowest dregs of the population, uncouth hooligans, gangsters, and cutthroats. As against them the Jews represent civilization here."

"Ah," he objected again, "but you must keep in mind that the government is neutral, must be impartial in this quarrel between Jews and Arabs. We are the watchdog . . . ."

"Neutral and impartial when it is a question of barbarism versus civilization, when it is a case of gangsters attacking peaceful, innocent citizens as those whose bodies I just saw piled up in a house in Hebron? . . . . Can one be neutral in a case like that? Isn't being neutral in such a case tantamount to taking sides, if one leaves the way open for more attacks, more murders of women and children?"

"By the way," Mr. Luke broke in, "are you a Jew?"

"No," I answered, "must one be a Jew to want fair play?"

"What then do you suggest?" he asked.

"I can suggest nothing," I replied, "I am only a newspaperman looking on . . . ."

"Quite so," he came back. "But you could be immensely useful to the government of Palestine. You represent a great liberal journal in America (the late New York *Evening World*), and we are very sensitive to American public opinion. I want you to have full facilities to probe the matter, to find out what really happened here, and to tell in what a grave predicament the government of Palestine was placed, caught between two fires, as it were . . . ."

"But I cannot say that," I returned. "You have just said the government is neutral. I can never hope to explain to the American newspaper reader that in the presence of an unprovoked attack by gangs of hoodlums on a peaceful community, the government looked on as an impartial spectator. Americans won't understand that. They expect something else from a government."

"What, for instance?"

"Well, a whiff of grapeshot, for instance—a few volleys fired into the air by the police. In the old days in this country, when the Turks were still the masters, this blood bath could not have happened. There is only one case on record of a Jew having been killed by an Arab mob in recent Turkish times. That happened at Jaffa. Jemal Pasha, the Governor of Syria at that time, summoned the Arab notables of Jaffa to Jerusalem, asked them who had been responsible and when they hemmed and hawed he had them taken out in the yard where they were given a beating with sticks on the flat of their feet. They soon revealed the name of the man who had incited the Jaffa crowd to riot. Then he sent them back with the provision that they walk all the way on their sore feet and bring back the culprit. They did. Jemal hanged him from the Jaffa Gate as an example and ordered all citizens to come and look what happened to a man who attacks his fellows and endangers the peace . . . . "

"But the Turks now, for heaven's sake, Mr. van Paassen, you wouldn't want me to behave like a Turk, would you?" laughed Mr. Luke.

"Not at all," I said, "but you can't be neutral either in a case like that."

"Well, no, but impartial we are . . . . "

"You call that impartiality when a man like Captain Caffaretta of Hebron comes in and calmly relates how he watched the mob invade a rabbi's home and slaughter twenty-seven persons there."

"There were only twenty-six killed," interrupted Mr. Luke.

"Yes, twenty-six adults and one baby of three months, that makes twenty-seven by my count . . . . Watch human beings being killed, he an officer in the British army, with a police guard at his beck and call and a service revolver in his pocket. One or two shots in the air by Caffaretta, and that mob would never have entered Rabbi Slonim's house."

"We arrested some of the attackers," said Mr. Luke.

"You arrested first and foremost in every case that I investigated," I said, "the Jews who successfully defended themselves. You arrested fifty Jews in Haifa at the moment they defended themselves heroically against the attack of a mob of some two thousand runners-amok . . . . Yesterday I saw a man being brought into Jerusalem by the mounted police and recognized in him an old settler from the neighborhood of Lifta, the owner of a small canning factory who has been in this country for more than fifty years—a real pioneer. I visited this man in jail. His name is Isaac Broza. He told me he was arrested after he had barricaded himself in his factory. And he has barricaded himself in his factory after the Arabs from Lifta had raped and massacred his old wife and his two daughters and had fired his house across the roadway. But the most scandalous thing," I continued, "the most incomprehensible of all is the treatment meted out to the two hundred Jewish young men who were sworn in as special constables a few hours after the rioting started here in Jerusalem. The government armed them and placed them on guard around Government House. Then, as soon as the tumult subsided somewhat and the first military reinforcements arrived from Malta, they were arrested on the charge of carrying weapons . . . . "

"Just a moment," interrupted the Acting High Commissioner, "does it not occur to you that we may have arrested these people, those in Haifa and Mr. Brozen and the supernumerary constables, too, so as to keep them in safety for a few days, until this thing blows over, that we placed them in protective custody? . . . . "

"That did occur to me indeed," said I, "but the circumstances of their arrest and custody do not warrant a belief that the government's sole object was their security . . . . "

"Why?"

"Because they were put in chains and in solitary confinement. The Jews of Haifa were chained up. I do not know about the supernumeraries here in Jerusalem. But Mr. Brozen was loaded down under chains when he was brought in yesterday from Lifta. When I visited him in his cell, in the company of his lawyer, he was still in chains, chains on his hands and chains on his feet, old Turkish chains at that . . . . "

"Well, you know, in days of excitement," he came back soothingly, "mistakes are unavoidable. You must not forget that this thing, this outbreak, took the government completely by surprise. We will rectify those mistakes as soon as order is restored. I will sign an order for Mr. Brozen's release at once . . . . "

"If this outbreak took the government completely by surprise as you say, Mr. Luke," I said to him, "how is it then that three weeks ago the Government warned the authorities of the Rothschild Hospital here in Jerusalem to have two hundred beds in readiness for an emergency?"

"Who told you so?" he shot back angrily.

"The directors of the Hospital, the doctors, Dr. Dantziger and Dr. Ticho," said I.

"Are you a French Canadian?" he blurted out.

"No, I am a Canadian citizen and became one by virtue of my enlistment in the Canadian army in the last war. But I am a Hollander by birth . . . . "

"Of the Jewish faith?"

"Neither of the Jewish faith nor of the Jewish race, I am a Hollander and nothing else. Why do you ask?"

"You seem to be extremely critical of what we have done here in Palestine. You seem to think that it is chiefly the Government that is in the wrong, do you not? Why do you call yourself a Hollander if you are a Canadian citizen?"

"I am a national of Canada, but I was born and bred in Holland, and all the naturalization papers, passports, and diplomatic hieroglyphics cannot change a man's blood and his ancestry. And as to unfairness, yes, I do think the British government—especially the Palestine administration—has not dealt fairly with the Jewish people."

"You are of the opinion, it appears, that the government should simply approve of everything the Jews undertake and do . . . . "

When I shook my head, he continued nevertheless: "But you overlook our obligations also to the Arabs. The mandate charges us specifically that 'nothing shall be undertaken,' " he quoted, " 'that shall interfere with the religious or civic rights of the existing population. . . .' "

"Did the Jews impinge upon these rights then?" I asked. "Are they impinging upon these rights by building roads, laying out plantations, building cities, hospitals, clinics, schools, banks, and all the other things? . . . ."

"I would not say that. But the Arabs are afraid, you see. They see new Jews arriving constantly. They know Jews are buying land. They are fearful of the future, afraid that they are going to be crowded out."

"You think that is a legitimate fear?" I asked.

"No, but what are we of the government to say when the Arabs point to the irrefutable evidence that Jews are really on the increase?"

"Why Mr. Luke, I should think that if you told the Arabs that the Jews intend the Arabs no harm, but to the contrary that they will do them good in every respect and that the coming of the Jews is in accordance with the provisions of the mandate which calls for a close settlement of Jews on the land, if the Arabs are told that this is Britain's will and the will of the civilized world and that no harm will come of it, I think the Arabs would abide by that decision. But if you, on the other hand, hesitate all the time, and act and talk as if you of the administration yourselves are not sure whether or not some injustice is being perpetrated on the Arabs, then you simply invite trouble. You get hooligans killing Jews here in Jerusalem and Hebron with the cry: 'The government is with us!' "

"I do not know whether the Arabs really used those words. It should be very unfortunate if they had,"

Mr. Luke interrupted me.

"And very revealing," I said.

"Dreadfully so," he laughed.

We talked a little more about the Mufti, the man who was proved to be the instigator of the disturbances then and in 1920 as well as later in 1936, and I asked why such a man was not arrested.

"We have no absolute proof," said Mr. Luke, "that he is behind it all."

"If I get you the proof, his own boastful statements to myself and to half a dozen other newspaper correspondents; if we get you a stenographic report of his sermons in the Mosque of Omar, will that help you?" I asked.

"It will help, no doubt," he replied, "but we can't expect you newspapermen to play the detective for us. Our evidence will have to come from our own agents."

"To be sure, you cannot let real neutrals interfere," I said.

And so the disturbances kept up: eucalyptus forests were cut down, isolated Jewish colonies attacked, individual Jews waylaid and ambushed, synagogues ransacked and set on fire. Bombs were tossed in the dark, buses fired upon, automobiles wrecked, farmhouses destroyed, young orange trees uprooted . . . until the Royal Commission of Inquiry arrived—no, not from Geneva, from the Permanent Mandates Commission of the League of Nations, a neutral, impartial body composed of nationals of half a dozen countries, but a Royal Commission from England, composed of functionaries of the Colonial Office.

Is it surprising that *les dessous de la pendule,* as Alphonse Daudet called the inner secret workings of governments—the real manipulations of the Palestine administration, its intrigues, its anti-Zionism, its anti-Semitism, its "Hebrew gold" and "poor Arab arguments," its wishing for outbreaks, and the next logical step: provoking outbreaks in order to coerce the recommendations and decisions by royal commissions and Parliament—were never mentioned at these official inquiries? The Palestine administration's role and the role of the Colonial Office were always "excluded from the terms of inquiry."

Once or twice the part played by the Palestine administration threatened to come to the world's attention. In the Permanent Mandates Commission in Geneva in 1938, for instance, when Dr. Hendrik van Rees, the member for Holland accused the British of being themselves "the conscious and deliberate fomentors of trouble in Palestine". Or when the member for Norway, Miss Dannewig, said in Geneva that not the Arabs but Britain herself was betraying the Jews and that to say otherwise was "insulting the intelligence of the members of the Permanent Mandates Commission". Or still again, when Professor Rappard, the member for Switzerland, a renowned authority on mandates, accused Britain of "turning the mandate upside down," in that she, who is obliged by international direction to create a Jewish majority in Palestine, repudiates what "in the Commission's opinion is the major element in the mandate."

It made no difference what the Permanent Mandates Commission said or did. The Colonial Office pursued its way unruffled. The guardian of the Holy Land would never account for or explain his own conduct. He always shifted responsibility for the terrors and upheavals to the shoulders of his wards: the Jews.

After every royal commission's investigation in Palestine the position of the Jews worsened. More restrictions were placed on their endeavors, more difficulties were placed in their way, more ordinances were set up to hamper them in their work.

Yet the Jews went on persevering. In spite of the Colonial Office, in the teeth of the local saboteurs of the Palestine administration, they went on building. When the disturbances of 1936 broke out, the Jewish community of Palestine adopted the policy of *havlaga,* meaning self-control, national self-discipline. The Jews decided with one common accord not to retaliate, not to resist evil with evil, not to kill innocent Arabs when such Arabs had been incited to kill Jews or destroy Jewish property. They could have wiped out the Arab terror bands in one month. Prominent Arabs actually came and prayed the Jewish defense organizations to do it, for hundreds of Arabs who refused to be intimidated by the terrorists in joining anti-Jewish depradations were slain or tortured by the hoodlum gangs, which in 1936 had the active support of the Nazi colonists of Waldheim-Bethlehem, Sarona, and Wilhelma, the old settlements of the Knights Templars in Palestine.

From the Royal Commission investigating the causes and extent of the turmoil of 1936-1939, which surpassed in length of time and in destruction of lives and property all previous disturbances, but which came to an abrupt end with the outbreak of the war—that is to say, when the Nazis and Fascists domiciled in Palestine were interned or departed—came the infamous White Paper of 1939. This document, which was forced through Parliament by the Chamberlain government in the course of the general appeasement policy pursued by Britain at the time, limited Jewish immigration to 75,000 for the next five years. After that the volume of Jewish immigration is up to the Arab majority in Palestine. This would mean, of course, a complete cessation of Jewish immigration.

But lest the Palestinian Arabs perchance come to realize that Jewish immigration is a far greater benefit to them than an evil, the Colonial Office brought in the rulers of foreign Arab countries—Ibn Sa'ud of the Hejaz, the King of Egypt, the Prime Minister of Iraq, and the Emir of Trans-Jordan—to declare in turn their unalterable opposition to Jewish immigration into Palestine.

With their support, gained in return for promises of establishing an Arab federation in which Jewish Palestine is to be included, the mighty British Empire dares to face the world with a determination to stabilize Jewish enterprise in Palestine on the present basis of accomplishment—in other words, to wreck and ruin and frustrate the movement to normalize the life and condition of the Jewish people of the world.

It was decided in 1939 that the final clauses of the White Paper relating to the total stoppage of Jewish immigration would go into effect in 1944.

But before that year dawned, something else had taken place in Palestine. The war had transformed the country into a great bulwark of British power and into one of the pivotal strategic areas of the global conflict.

CHAPTER 4

THE BEST-KEPT SECRET OF THE WAR

IN the summer of 1942 the situation of the British army in Egypt was little short of desperate. It had lost more than half its man power and the better part of its mechanized equipment to Marshal Rommel, who was boasting openly that before six weeks were over he would be sleeping peacefully in the royal suite of Shepheard's Hotel in Cairo. And why shouldn't the German commander have been confident that final victory lay within his grasp? Fortune's every augury pointed in his direction. The Egyptian plum seemed ripe for the picking. Rommel merely needed to stretch out his hand, so to speak, and the British Empire's life line was about to snap in twain. Hadn't he completely nullified Wavell's earlier conquests and hadn't he thrown Auchinleck out of Libya in one brilliantly executed stroke? One more such move, just one more such resounding blow by his perfectly co-ordinated Afrika Korps could scarcely have any other effect than sending the British reeling across the Canal or scurrying for shelter into the trackless wastelands of the Sudan....

Rommel's scout planes were over the Nile Valley

and over Suez within twenty minutes after rising from the desert bases. Through his field glasses the Marshal could see the needle point of Pompey's Column in the heart of Alexandria as clearly as the minarets of the Mohammed Ali mosque, all of it less than seventy miles, less than a couple of good days' drive distant. He had ample cause to be hopeful of terminating the issue by one last determined push. The goal was in sight. He said one day that he was playing a game of cat and mouse with Montgomery and that he could make the kill whenever he chose to do so.

The Mediterranean sea lanes were dominated by land-based Axis bombers so that ships carrying reinforcements in men and material to Rommel's opponents were compelled to make the enormous detour around the Cape of Good Hope. Even so, less than half the new equipment from Britain and America succeeded in getting through the packs of submarines hunting off the South Atlantic coasts. On the other hand, Rommel's own, much shorter line of supply remained intact. By way of Italy, Sicily, Tunis, and Benghazi a steady stream of men, tanks, planes, oil, and munitions was pouring into his camp. What could the British do but fall back when the already victorious Italo-German machine was geared up once more and came rumbling forward?

Britain's plight seemed indeed irremediable. Her great cathedrals had crumbled under the mass assaults of the Luftwaffe. The one port after the other—Portsmouth, Southampton, Liverpool, Newcastle, Plymouth—had flared up in the night, turned into piles of smoking dust and ashes in the morning. Across the Channel, twenty-six miles distant, the German invasion barges lined the shores from Dieppe to Calais and from Ostend to Walcheren. The guns and the tanks that were to beat England into subjection were wheeling up along a hundred parallel roads leading to the Channel coast. Nazis stood massed in Holland, Denmark, France, Norway, and Belgium. America's supplies were still a mere trickle. What equipment could Britain spare for her armies in Egypt when her production had not made up for the losses sustained in the disaster of Dunkirk and when the Russian ally, now pressed back to the Volga line by Rundstedt's sledge-hammer blows, clamored in desperation for help? Singapore had fallen. The Japanese army was pouring into Siam and Burma. Corregidor had surrendered. The valiant Dutch navy had sunk beneath the waves. India was menaced, and Australia angrily insisted on the recall of her divisions and air force stationed in the Middle East.

While Rommel put the finishing touches to his preparations for the last battle, the bells of Berlin's churches were ringing in anticipation of the fall of Cairo and Jerusalem. The Duce ordered the Italian cities "smothered" under flags and bunting in joyous celebration over Britain's awful predicament. Young Blackshirts danced in the Piazza del Popolo. Not a word about the inhumanity and cruelty of aerial attacks came from Vatican City when it was a case of Protestant England being bombed to smithereens. Francisco Franco had his Phalangists demonstrate under the windows of the British ambassador in Madrid and shout their master's disdain for the rocking Empire of Britain with the cry: "Gibraltar! Gibraltar!" Over in

Paris and Vichy *ces messieurs* of the French General Staff and the new Nazi order, the Déats, Pétains, and Peyroutons, who had brought in the German armies to put an end to what Weygand once called "all this democratic nonsense," rubbed their hands in satisfied glee as the former comrade, now half blinded by sweat and blood, staggered on his feet. Their prophecies of England's fall were at last nearing fulfillment. The giant was groggy at last. The world's policeman was now obviously himself in a trap. They could hear his labored breathing. In Madrid, Vatican City, Tokyo, and Berlin they listened for the death rattle to begin.

In whatever direction they looked in that critical hour the British people saw little better than enmity, defiance, animosity, or sullen indifference to their fate. Isolationist moralists proclaimed from their platforms and in their press that now the transgressions of the fathers were being visited on the children, that England was getting her just deserts under the law of divine retribution. All the Empire's sins—in Ireland, in India, in South Africa—were held up to the public gaze. There were men in America who calmly envisaged the destruction of the British Empire as if that event would not have shaken the very ground from under their own feet.

It would have been a black day indeed for the human race, as Santayana once said, if scientific blackguards, conspirators, churls, and fanatics had managed to supplant Great Britain!

England stood alone. The world held its breath as a great empire seemed about to descend into the tomb. What is the mysterious power that sustained the

British people as they faced a seemingly inescapable disaster and the gates of doom with a stoicism and silent equanimity reminiscent of the *gravitas* of the Romans of the golden age? England stood against the world: flesh and blood against fire and steel. Yet not a whimper came from her lips. The English soldier, remarked a man who saw him, grows silent and faces death as if he shares a secret with his Maker. England was silent in the summer of 1942. It was her greatest hour!

But the crux of the desperate situation lay in the Near East. And there an ominous and sinister stillness prevailed at the British army's back. Young King Farouk of Egypt and his ministers had refused to lift one little finger in the defense of their invaded country. We know from diplomats who were in Cairo at the time that Nazi victories were commented upon with hopeful smiles and an exchange of meaningful glances in the palaces on the Nile. In Palestine the effendis (landed aristocrats) were telling the fellahin (peasants): "Now go and sell your land to the Jews and be quick about it, for in a month Hitler will be in Jerusalem, and you will not only have your land back but everything the Jews possess! Let the knives be sharpened! The great day is about to dawn! The Jews' protector is beaten!"

Twenty-four hours of the day the radio stations of Bari, Palermo, and Berlin were screaming the promises of Mussolini—the "Sword of Islam"—in the Arabic language: "Loot immeasurable, death to the English and Jews!" The two honored guests of Adolf Hitler—the ex-Mufti of Jerusalem, Haj Amin Effendi El Husseini, and the ex-Premier of Iraq, Rashid Ali-el-Gailani Bey, announced their speedy return and a settlement of old scores. In the mosques of Baghdad, Cairo, Amman, Jerusalem, Damascus, and Mosul, a careful investigation revealed, ulemas and muftis time and again working up the believers to a frenzy of excitement by pointing to the nearness of the hour of quittance in blood with all the *roumis* and infidels.

Let us have done with the preposterous myth of the Arabs eagerly waiting for an opportunity to rush to Britain's aid! The truth is that the so-called Arab world —that is to say, the princes, potentates, imams, mullahs, and emirs—were yearning to stab Britain in the back. King Ibn Sa'ud, who was suddenly to declare himself an ally of Britain and America, had not a word to say and could not spare a single trooper, camel, or donkey when Rommel stood at Alamein. The Imam of Yemen had to be watched constantly. Both princes received subsidies from the British government to keep the peace. In Iraq, whose attempt to oust the British garrison and hand over the Mosul oil fields to Germany had just been defeated, the instructors and students at the War College in Baghdad, as one British Intelligence officer relates, were still regretfully lamenting, a year after the event, their failure to establish contact with Nazi paratroops at the time of the revolt.

Had one German division set foot on the Syrian or Palestinian shores, the British command in Egypt would have had a conflagration at its back running from the Persian border to the Hadhramaut and Aden. As it was, almost as many British troops were required to patrol the doubtful Arab areas and cities, especially in Egypt, as there were actually facing Rommel's Afrika Korps in the Libyan desert . . . .

Not only in the Near East were the prospects utterly gloomy; the skies were darkening everywhere. Just as Rommel was stealing up and crouching for the last leap at Suez to cut the British Empire's jugular vein, the Red army received the full impact of Hitler's concentrated attack in the Donetz Basin. Timoshenko was forced to abandon Russia's most powerful industrial area, the richest coal mines, and the strategic railroads of the Caucasian transportation system. Sevastopol fell. Novorossisk followed and the entire Kuban district. The Nazis were pouring from the Crimea into Caucasia. They were marching around the shores of the Black Sea. In a few weeks' time they stood in the passes of the Caucasus Mountains within sight of Russia's refineries and oil fields. They were on the point of seizing those raw materials which would have enabled them to go on fighting indefinitely.

Men are apt to forget, now that the design has been gloriously frustrated by General Montgomery, that the pretentious objective of the Axis in the Near East was to cut the British Empire in twain by bringing about a juncture of German-Italian naval power with Japan's via the Mediterranean, the Suez Canal, the Red Sea, and the Indian Ocean. Every German strategist of importance attributed Germany's defeat in the war of 1914-1918 to the fact that British naval power had succeeded in throwing an impassable cordon of steel around the Reich, thus depriving it of the raw materials with which modern wars are fought. If the Germans in the First World War had been able to reach and draw upon the stocks of Asia and Africa, the analysts of the defeat were convinced, an entirely different story would have been told in the Hall of Mirrors at Versailles in 1920.

The supreme and costly error in the calculations of the Kaiser's grand strategists, it was discovered, lay in the neglect of the German High Command to support the Turks adequately in their drive on Suez in 1915 and in not pursuing the Near Eastern objective by a campaign of more determination and magnitude. Egypt and Palestine were the British Empire's most vulnerable spot, the solar plexus of its global defense system. There, in the neighborhood of the Canal which linked the British motherland with the reservoirs of wealth and man power in the East and the Antipodes, where "the thin red line" was thinnest and most dangerously exposed, the blow could have been struck that would have doubled the Empire back on itself and made the infliction of a mortal wound possible.

The fundamental error was, Generals Ludendorff and Buat admit in their memoirs, that too many divisions were diverted to the Russian front instead of being sent to the support of Turkey in Asia Minor and in the Arabian Peninsula. The Turks, after a brief offensive show, were neglected and allowed to shift for themselves. They, too, turned their attention towards Russia, confining themselves to defensive operations in the Near East, whereas they should have done the opposite: attack in Egypt and be content with defense in the Caucasus. Russia, after Hindenburg's crushing victories at Tannenberg, the Masurian Lakes, and in the Carpathians, was disintegrating rapidly and could have been held in check with a relatively small effort. On the Eastern Front huge German forces lay immobilized for years while victory's prize lay in another direction. German strategy mistook that direction for a mere diversion and treated the war in the Near East as a side show, whereas Germany's major efforts should have been concentrated on Arabia and Egypt, on driving the British from the eastern entrance to the Mediterranean.

But if that mistake had been made in 1914-1918, it was not to be repeated in 1939. If the German armies failed to reach the wealth of Asia and Africa once, they were to do better next time. New weapons, new tactics, a new spirit, a bold strategy of global dimensions would, so it was figured, overcome the obstacles that had formerly barred the Reich from its place in the sun. Air power was the new weapon. The new Germany of Hitler would wing itself over the walls of steel that British naval power might throw up again around the Reich or around a German-controlled Continent.

In the new conception of strategy, elaborated under the Weimar Republic which left the old General Staff undisturbed, and adopted as the official fundamental doctrine of war upon Adolf Hitler's accession to power, the conquest of the European *Herzland* was a preliminary move on the road to world power. This Britain could not prevent and might even consent to if Germany's secret objective were presented as the colonization of Bolshevik Russia. With the rapid collapse of France, it turned out a relatively easy undertaking. Europe fell under the German heel. But the real objectives in the new war the Germans unleashed lay beyond. Africa and Asia were the essence of the Greater Germany's *Lebensraum*. These two continents, which were renamed Eurafrica and Eurasia in the new geopolitical strategy—that is, mere economic adjuncts of the German-controlled European "heartland"—were the principal goals Germany aimed at in her new bid for world power.

By occupying or gaining control of the principal bases and sea lanes of Africa it was clear that the South American continent, too, would inevitably fall within the orbit or at least be drawn within the sphere of influence of the Third Reich. Without gaining control of the Mediterranean, which was, in the conception of Dr. Haushofer, the theoretician of the new geopolitical school, not a barrier of water separating Europe from Africa but the link binding these two continents together, a mere inland lake of Eurafrica, and without control of the South Atlantic sea lanes and the resources of Africa and western Asia, the conquest of Europe would remain a sterile victory, an incompleted effort in that the Reich could never hope to fulfill its ambition of becoming the all-dominating industrial and the one unchallengeable military power in the universe.

Only a few men in the democratic countries were aware of how near Germany came to translating that fabulous dream into reality. President Roosevelt was one who early grasped the stupendous magnitude and significance of the plot. Yet the very fantastic nature of the project, the gigantic scope of the German imagination, would have turned a description or analysis of it on his part into a cause for raillery and opposition to his preparations for defense. The vast German scheme, even when it was being worked out with startling precision before their eyes, escaped the comprehension of his parochial-minded opponents, who sought refuge in ridicule and irreverent demagogic wisecracks. The President had to content himself with the oft-reiterated warning that America was in grave danger, without specifying the exact nature and size of that danger.

Unprepared and almost defenseless, America was in a fair way of being taken in a gigantic, hemispheric pincer movement, with the European-African coast on the one side and the coast of Asia on the other representing the two arms of the nutcracker. A bridge of air power that could be thrown across the Atlantic between Dakar in Vichy's hands and the bulge of Brazil further emphasized the perilous position of America. By stirring up trouble in South America and massing air power there, the Germans planned to throw America back on a defense of the continental United States only, avoiding a frontal attack on the defenses of the Panama Canal by outflanking them, as they had outflanked the Maginot Line, and to launch their invasion of America by way of the Gulf of Mexico and the Mississippi Valley. Once or twice the staggering military plot that Germany hatched against the United States was revealed by certain periodicals and analysts, only to be dismissed by the public as too fanciful and preposterous and too farfetched. Isolationists ridiculed the idea that America was in any danger at all.

How near Germany's geopoliticians and military strategists came to fulfilling that bewildering dream of theirs men will probably not realize until the secret history of the war is written. For two years the German armies moved irresistibly and with inexorable determination towards their goal. The Czechoslovakian fortress jutting into the Reich's most vulnerable flank was removed by an adroit political maneuver. Tory Britain was successfully hoodwinked in believing that with the removal of that bulwark Germany's road to the Russian East was cleared. Poland was eliminated in one stroke. France fell. The gallant armies of Yugoslavia and Hellas were overwhelmed. The brown legions surged victoriously through the passes of the Thermopylae and filled the peaceful Peloponnesus with the metallic clatter of their assault cars. In a month they were in Crete, halfway across the inland lake of the Mediterranean.

In the slim finger tips that Hellas dips into the sea in Laconia, nature herself seemed to point the way to the gray hosts that poured through the Balkans for the shores of the old world sea. With the arrival of the German army in Marseille shortly before, the two main roads to Africa had been cleared. The great objective was in sight. Within a month the Germans had occupied the Cyclades and thus extended a bridge to the isle of Rhodes where Mussolini had been teaching landing tactics to his Army of Asia. From Mount Carmel in the Holy Land, men were daily scanning the skies expecting to see landing parties approach from Cyprus across the bay. Messerschmitts and Junkers roared over Nicosia bound for Beyrouth and Damascus, where German technicians were helping the Vichy French to lay out new airdromes. The ground for the invasion of the Syrian coast was prepared. Discerning spirits everywhere in the world recognized that one of the decisive hours in history was about to strike.

In the border areas of Europe and Asia, as much as on the borders of Asia and Africa, it was for the German High Command, merely a question of one more determined push in June, 1942, of delivering one more crippling blow: the *coup de grâce*. Like the arms of a monstrous octopus—Rommel on one side, Von Rundstedt on the other—the German armies were enveloping the whole Near and Middle East in a seemingly irresistible nutcracker movement . . . .

At that moment, it may now be revealed, the British High Command thought the game was up. Winston Churchill personally telegraphed the military authorities in the Near East to construct a set of bridges across the Tigris and Euphrates Rivers in Iraq. His order had precedence over all other pending tactical plans or defense projects. London feared a simultaneous German break-through on the Nile at Alexandria, on the Volga at Stalingrad, and in the Caucasian Mountains near Baku and Tifflis. If that had happened, it must be obvious that the British armies in the Near East would have been cut off. In the event of a break-through and of the two German armies converging towards each other, through Syria from the north and through Sinai from the south, there was no other solution but to retreat, if the British armies were not to be caught between two fires and annihilated. It was therefore planned, if the break-through occurred, to abandon the Suez Canal, Sinai, Palestine, Lebanon, Trans-Jordan, Syria, and Iraq and to attempt a retreat in the general direction of Persia. To facilitate the retreat and save as much matériel as possible, bridges were to be thrown over the two great Mesopotamian rivers.

But where could they get the steel and the material and the technicians and the crews of workers to build those bridges? The British army in Egypt could not spare a single engineer. Material was not known to be any nearer than Leeds or Birmingham . . . . After a few days' deliberation, British military commanders in Egypt and Syria were on the point of notifying London that the plan could not be carried out and that another scheme for the retreat must be provided at once.

It was at this moment that a Jew from Czechoslovakia, a businessman who lives in Palestine, happened to come into the office of the British military commander in Jerusalem. The Jew had come to get the commander's signature to an order of sheet metal which his small plant manufactures. As he sat in the antechamber of the general-headquarters building in the old Russian Compound, he saw several officers come out of the conference room. When he was summoned to come in himself, he found the harassed-faced commander bent over a map of Iraq and sighing disconsolately: "It can't be done! My God, I don't see how it can be done!"

A reliable and highly placed informant in the Middle East told me that the Jew hesitatingly inquired what the trouble was, whereupon a dialogue along the following lines developed between the two men. The commander explained first the predicament he was in: the immediate vital need of bridges in Mesopotamia, the lack of experts and material . . . . "I am afraid we are caught," he said, biting his fingernails.

"No," said the Jew, "we are not . . . . "

"If those bridges aren't built we are definitely caught," repeated the commander, as he looked his visitor straight in the eyes, with defiance almost. "The British armies are in a trap. There is no other word for it."

"If you let me, I will build those bridges for you," said the Jew.

The Englishman looked up. His jaw dropped. "You?" he said. "Do you know what is involved? Do you realize that this is an engineering feat of the first magnitude, the Euphrates, mud flats, miles wide . . . . Did you ever build a bridge before?"

"Yes," said the Jew, "I built two bridges, two of the largest in Europe, one over the Danube and one over the Moldau in Czechoslovakia . . . . I also built the subway in Berlin . . . . "

"Great God, man!" exclaimed the Englishman, jumping from his chair and seizing the Jew by the arm, "Do you know what you are saying?" And again he explained the fearful plight of the British armies.

"Yes," said the Jew, "we are all in a terrible fix, if there is a break-through . . . . Our old people, about a third of Palestine's Jewish population, carry poison in their pockets . . . ready to die. Our young people will fight to the last, to the bitter end, of course. But we all know that the civilians cannot be evacuated . . . . Still, the army must be saved . . . . If you give the word, I will build the bridges."

"Where will you get the men?"

"Leave it to me!"

"Where will you get the material?"

"Leave it to me!" came back the Jew again. "Just let me go to Iraq and have one good look, and I will tell you in a few days how soon the job can be done . . . ."

It was left to him and the bridges were built.

A month before British forces invaded Syria, where the Vichy regime was smoothing the way for a German invasion, twelve Jewish boys answered the secret call of the British military authorities in Jerusalem to undertake the hazardous job of blowing up the oil installations in the port of Tripoli in Syria, the terminus of the French branch of the Mosul pipe line. British soldiers could not be sent, as France and Britain were not at war. Moreover, the raiders had to be men who understood both French and Arabic, as circumstances might force them to spend a considerable time in Syria before they would have an opportunity of approaching the refineries, which were, of course, under heavy guard.

The Jewish boys who volunteered were told that not only must they not expect a reward or recognition in the event of a successful conclusion of the raid, but that Britain would have to repudiate and even denounce them as spies and saboteurs if they were caught. If they should be captured and put to torture, they were told they were not to divulge who had sent them on their mission of destruction. They were commanded to take the secret with them into death . . . .

Under command of a British officer, the twelve volunteers worked out their own plan of attack and one night dashed off from Haifa in a speedboat crammed with high explosives. They came undetected, through the crowded harbor of Tripoli, got ashore and found their way to the oil installations, overpowered some French guards, and started their work of demolition. The alarm was given at once, and they were surrounded while still busy setting fires and tossing their bombs around. There was no chance of getting back to their boat. It is not known whether they were formally executed or cut down on the spot. After the British occupation of Syria remnants of their bodies were found in a shallow grave near the shore and identified by Hebrew lettering on the clothing. Then only were their parents informed of their death and sacrifice by the Palestine administration. And their relatives were informed at the same time not to expect any premiums or pension money, as the boys had not been regular British soldiers.

When the First Brigade of the Free French, mainly composed of foreign Legionnaires, after marching north through the Sahara from Fort Lamy on Lake Chad, was trapped by Rommel at Bir Hacheim,* in Libya, in June, 1942, and there, clamping itself to the bare rocky soil, performed the miracle of another Verdun by holding the line for Montgomery for a whole month during the most critical stage of the battle of Libya, repelling daily mass assaults by tanks and undergoing almost hourly bombardment by huge flocks of Stukas, a company of Jewish engineers belonging to the King's West African Rifles went through the same experience at Mechili, fifty miles to the east. The Jewish military

**L'Epopée de Bir Hakim,* by Jean-Pierre Besnard, *Les Oeuvres Nouvelles,* Editions de la Maison Française, Inc. New York.

epic is related in the communiqués of General Koenig of the Free French. They were spotted by German scout planes as they were laying down a mine field which was to be a bar to possible attempts by Rommel to turn the flank of the Eighth Army as it stood with its back to Alamein. The Nazi observers caught sight of the Jewish engineers working in the open desert, swooped down to strafe them, and then flew off. That was on June 1. The next two days the whole plain around Mechili was blotted out in a whirling, yellow sandstorm which made the working party invisible from on high. The Jews could hear the Messerschmitts and Junkers roaring above the clouds of dust and apparently looking for them. Visibility on the ground was down to absolute zero.

Mechili is a stretch of flint-strewn sand with lava outcroppings, bordered on the east by the first undulating sand dunes of the Sahara. It is not an oasis, as some dispatches have made it out to be. Here and there you find a clump of burnt shrubbery, the color of ocher, but no trees, no shade, nothing but scorched desolation—sand and rocks. There is a well there, however, at which I stopped for a night in 1934 on my way back from Fort Lamy whither I had accompanied the scientific expedition of Professor Charles Perrault that was investigating the causes of the unexplained phenomenom of the sinking of the level of Lake Chad.*

The Jewish company's task in that area was to lay down a mine field in the shape of a quadrangle on a surface four miles by three. On the borders of the field

---

**Days of Our Years,* Garden City Press, New York, page 252 *et seq.*

the mines were buried two yards apart; farther inside the field they were more widely spaced. Those mines are the size of a soup plate and consist of a metal box filled with an explosive. They are placed in a hole and then covered up. In a few days' time the sand levels the earth and no visible trace is left of the deadly machine. Such mines are designed to halt trucks weighing over five hundred pounds, tanks, and artillery. They do not explode when a man walks over them.

The engineers had scarcely begun their work, tracing the outline of the mine field by driving down stakes and connecting them with barbed wire when the Germans spotted them. As soon as the sandstorm subsided sixty heavy bombers paid them a visit and blew up half of their trucks. When the Stukas returned the following morning and twice more in the course of the afternoon of June 4, the officer in charge of the mine-laying operations, Major Felix Liebman, a citizen of Tel Aviv, heliographed the nearest British post for some antiaircraft guns. The answer came back that ten antiaircraft guns would be sent immediately, plus some antitank guns, as the Jews must be prepared to face a land assault any moment. British scout planes had observed a column of Nazi tanks moving in the direction of Mechili. With that message came word from the Commander in Chief himself, General Sir Bernard Montgomery, to finish laying the mines and to hold the field at all costs. The very fact that the Germans were paying so much attention to Mechili showed the importance they attached to the place. Reinforcements, too, were promised.

But these never arrived. In three days' time Mechili

was surrounded on three sides by a ring of enemy tanks, both German and Italian, and the engineering garrison was cut off from all contact with the outside world. Before attacking, the German brigadier in charge of operations sent a tank with a white flag and the message to hoist the signal of surrender. That happened on the morning of June 10. Major Liebman said to the German officer who brought the message: "We have no white flag! All we have is the banner. This we are going to fly. It's the blue flag of Zion! . . . ."

"Sie sind Jude" (You are a Jew), said the German in surprise, clicked his heels, saluted, and walked off.

Six hours later the tanks came rumbling towards the position, sixty in one column and twenty-five each in two others. A flock of Stukas which appeared simultaneously was forced to drop its bombs prematurely when it was attacked by British Kittyhawks. But the tanks rumbled on. The Jews held their fire until the first metal masters reached the barbed wire stakes. Then they let go. Two tanks blew up when they struck mines; nineteen were hit by antitank fire. One Jewish sergeant alone accounted for seven of them.

Meeting with so much unexpected resistance, the main tank column, which belonged to the Italian Ariete Division, halted, signaled to the others, and started to withdraw. This was the moment for which Major Liebman had prepared. Sixty of his men who had been hidden in dugouts near the extremities of the mine field rushed out when the Italians turned tail and bombarded the retreating tanks in their vulnerable rear with hand grenades, bottles filled with gasoline, and tommy guns. Some of the Jews jumped on the back of the tanks, firing their revolvers into the lookout slits and gun holes. In this way five more enemy machines were accounted for. As the last Italian caterpillars moved off British fliers swooped low over the mine field and waved their hands at the Jews.

The following day the Germans subjected Mechili to a merciless aerial bombardment. They repeated the assault twice a day. They had apparently decided to reduce the position by air attack, wipe out the little garrison, and clear a path for their tanks around the leftmost extremities of Montgomery's flank. For seven days the bombs rained down, turning the mine field into a wailing hell of steel in which it did not seem possible for human nerves or human life itself to endure. One squadron of Stukas had not passed over and dumped its ghastly load before the next one winged into sight. They dive-bombed the trucks, the dugouts, the guncrew. They churned and plowed up the mine field, filled it with craters two stories deep until Mechili was an inferno of boiling red-hot, iron missiles in an inferno of blistering desert heat. Still the Jews held out.

On June 20 the tanks returned en masse. Upon their approach the weary, hungry defenders clambered out of their foxholes and manned the guns. Again they repelled the assault of the Ariete Division while an aerial battle went on above their heads and British fighters drove off the Stukas come to administer the *coup de grâce* to the garrison. But at the end of that day, although forty-one smoking tanks testified to the deadly accuracy of the Jewish gunners, only ninety men were left out of Major Liebman's original five hundred.

Ten more days passed. Each day the Italians renewed the attack, raking the mine field with a murderous fire, getting nearer and nearer to the central position with each onslaught. Then the water well was hit and stove in by a well-placed Stuka bomb, and the agony commenced. From that day on the men were reduced to a daily pint of water from the tin cans dropped by the R.A.F.

Major Liebman had banded his men closely together around the deep, central dugout. Here they were going to make their last stand. Outside lay the bodies of their comrades which the Stukas would not leave buried but plowed up and sent up in the air in a ghastly, maddening dance of rotting flesh and bones. Two men went out of their minds on June 25, two more the next day. Three men rushed off shrieking into the desert. The bombardment continued. With the summer heat mounting every day and the scorching wind blowing up clouds of dust, the men's thirst grew unbearable. Some drank gasoline and perished. Other drank their own urine and went mad with the pangs of greater thirst afterwards. Nobody spoke a word those last days. Rifles grew burning hot in the men's hands. The dive bombers sent up geysers of sand around them.

Forty-five men were left on July 1, a handful of unrecognizable scarecrows, scarcely human in appearance, unkempt, haggard, covered with grime, emaciated, some stark naked, having had the clothes blown off by the concussion of five-hundred-pound bombs exploding near by.

On July 2, they faced their last assault at six in the morning, lost two more men, and put the Italians to flight once more. At ten o'clock, a lookout man, who could scarcely speak, his tongue cleaving to the roof of his parched mouth, warned Major Liebman that a column of trucks was approaching, led by an automobile bearing the tricolor of France. The commander, although wounded in the head and in the groin, staggered to his feet and waved the answering signal.

The French troopers approached. They were the remnant of the Free French from Bir Hacheim who had received orders to withdraw the night before. The Jews stumbled into the open, looking like so many tortured ghosts. General Koenig of the Free French walked up to Major Liebman and embraced him. "*Vous avez tenu bon, jusqu'au bout,* you held out till the end," he said. Then tears choked the Frenchman's voice.

The Jews were given water, and the Major informed them that they were to accompany the French column. The siege was over. French soldiers were loading the remaining equipment on trucks. A flock of Spitfires sailed by, the pilots waving their helmets.

One Jewish soldier took down the blue and white flag of Zion, rolled it up, and was about to place it in its holder when General Koenig saw him and asked him a question.

"We are not permitted to fly that flag," explained Major Liebman. "It's against regulations . . . . "

"*Pardon,*" said General Koenig, "I am in command here. *Je m'en fous pas mal des régulations,* I don't care a damn about regulations . . . . That flag goes on my car in front, next to the tricolor. That's where it belongs. *Nous sommes victorieux, tous les deux,* we have both come through victoriously!" And turning to his men, the French officer shouted: *"Légionnaires! Le drapeau juif! Salut!* Legionnaires, the Jewish flag! Salute! . . . ."

"Just as soon as we start the big push," said General Montgomery, "I want to create a little diversion behind Rommel's lines. I would like to take one of his supply depots on the Libyan coast. I had thought of the town of Bardia, that is the nearest to us. I do not think we could hold it for any length of time, for the place is strongly held by an entire division of Italians who have German artillery support. But holding on to Bardia is not the first essential at the present stage of the game, although permanent seizure would, of course, be a big help. For the moment I would be satisfied with raising hell there for a few hours, blowing up the munition dumps and the petrol supply which is stored in the caves near the shore, wrecking the tank and aviation repair shops, and ruining the harbor. What do you say? Do you think it can be done?"

The words were addressed to Commander Osterman-Averni, chief of a Jewish "suicide-task force" from Palestine serving with the Eighth Army. Commander Osterman-Averni has told the story in the Hebrew daily newspaper *Hamashkif,* which is published in Jerusalem and I have verified it from other sources.

"Three days after General Montgomery called us to headquarters," he writes, "we were inside Bardia. But we did not go alone. I mean my task force was joined by another Jewish suicide commando. Together we went in. I wouldn't be surprised if the Italians of the

Bardia garrison, who are now nearly all prisoners of war, were still trying to solve the riddle of how we got there. Actually, the answer to their puzzle is extremely simple.

"We were put aboard two destroyers in the late afternoon. As usual, the men were not told in advance of their destination. They imagined that our task would be one of those routine 'behind-the-line' actions: the demolition of a bridge, the destruction of a water supply line, or some similar task. They did not have any particular reason to devote much speculative thinking to the task ahead . . . .

"When we were nearing land, I told the men that we were going to land at Bardia and that, if possible, the town was to be taken in a general assault at dawn. I told them it would not be an easy job and emphasized that strict discipline and group spirit alone could insure success. I said it was the most important task entrusted to us thus far and that the honor of the Palestinian 'suicide forces' was at stake. 'If we come through,' I said, 'I am authorized to promise each of you an additional stripe.'

" 'If the honor of the force is at stake, we will be in Bardia tomorrow morning,' spoke up a sergeant.

"We approached the Libyan shore in Stygian darkness," Commander Osterman-Averni goes on to write. "The destroyers scarcely moved as the rope ladders were let down by which we slid into the rowboats that were to set us ashore. These boats advanced stealthily. Nobody spoke. The oars barely skimmed the waters. Not a speck of light showed in Bardia. In fact, we could not even see the coast. The scraping of the boats on the rocks was the first intimation we had that we had reached our destination.

"In deepest silence we waded ashore. A patrol was sent forward immediately. We waited an hour. From the distance came the thunderous roll of our artillery. We could hear the metallic steps of the Italian sentries on the quays. When the patrol returned, reporting that they had established the exact position of the spot where we had landed, the British sailors from the warships whispered 'good luck' and dipped their oars in the water. Then we were on our own. The last contact with the Eighth Army had been broken . . . .

"We were eighty-five men in all. We had ten machine guns which required the services of twenty men. The rest of us were armed with tommy guns and knuckle-duster daggers. We had one signalman and one medical man with us.

"We advanced in the dark. Some scouts went ahead, their daggers ready for instant action. We lost all sense of time. Every minute was like an eternity. We reached the road at last, stopped, and lay down to await the report of the scouts. In this neighborhood we knew there should be an Italian guard post. Our machine guns were at the ready, to meet all eventualities. But we also knew that we must not fire yet, for it would have betrayed our presence. And that, considering that we were eighty-five men against a division, would have meant our certain annihilation.

"Forty-five minutes we waited. I was growing anxious about our scouts. Then the sound of a shrill low whistle came to us in the dark. A scout came running back. The Italian post had been taken. The scouts had carried out a 'silent job'. As we stepped up I noticed several bodies on the ground. I could not tell whether they were corpses or just stunned or tied. Nor did I care.

"Ten lorries then rumbled past over the road. They did not even dream of stopping to examine the isolated guard position. Then we advanced, the machine guns in front. Dawn was breaking. As we marched along the road into Bardia, another caravan of trucks passed by, going in the same direction as we. Our main danger was that the drivers might offer us a ride. But to our luck, all the trucks were heavily loaded, and nobody bothered with us hitchhikers. It did not seem to enter the drivers' heads that an enemy party was marching along right in their midst.

"So we kept walking until we saw our chance and made off across a field. Several blockhouses were dealt with. Their small garrisons were given 'silent treatment.' We managed to advance to a well-built concrete blockhouse. The garrison was still asleep. We quickly dispatched all of them while they lay on their cots. Then we made ourselves comfortable, set up our machine guns, and waited.

"When an hour after dawn a deafening roar heralded the artillery barrage of our own guns, indicating that the British 'push' was on, we heard the signals all around us, calling the blockhouse garrisons to the defense. Italian soldiers streamed out of them and ran forward. And we opened fire on them.

"One officer, thinking that his men were being fired on by mistake, shouted at us, but we continued to fire. After a while it dawned on the Italians that something was wrong. Two companies appeared, cautiously approaching our position. When they were quite near, we hurled a well-aimed mass of hand grenades at them. They dispersed in panic.

"Fire was opened at us by their machine guns from all directions, but it was ineffective since we were in a good position of concealment. They got their artillery to open fire on us too, but their fire was misdirected, since their own men were all over and they could not easily pick out as a special target the one blockhouse we occupied.

"But now shells of our own guns began falling dangerously near. Realizing that we were no longer an unknown quantity, we flashed signals to our own observers overhead and in the general direction of where our main forces might be. After an hour or so we knew that we had been seen, for the shells of the British artillery outside Bardia began giving our blockhouse a wide clearance.

"We then redoubled our fire on the Italian rear, and their officers, believing themselves surrounded by superior forces, hoisted white flags on the blockhouses. The town of Bardia was ours. But we did not leave our blockhouse except to occupy a few more of the neighboring pillboxes, where we manned the captured Italian machine guns. We could not very well show the enemy how few we were, for in that case he might well have regretted his surrender and turned on us. By midafternoon the first wave of British infantry and the motorized units moved into Bardia without firing a shot.

"We did not raise hell in Bardia, it is true. We did

better: we captured everything intact and nine thousand enemy prisoners to boot. . . . "

In June, 1942, when Rommel dropped that contemptuous remark about playing a mere game of cat and mouse with the British Eighth Army and the whole civilized world wondered why he delayed giving the signal that would send his Panzer divisions on what appeared in all likelihood to be their last single day's drive into the Nile Valley and to the shores of the Suez Canal, General Bernard Montgomery was as much in a quandary as anybody else. We know now that the Nazi commander was purposely biding his time, perfecting his plans to the last minute detail, collecting such enormous supplies that when the hour of attack finally struck his last dash would be one glorious, irresistible climax to the long desert campaign—an assault so powerful that British resistance would melt at the first onset and turn into an unparalleled debacle. But this was not known at British headquarters. At least, no one there could imagine the real cause of delaying the blow that might well prove fatal. Nobody knew exactly what trick Rommel had up his sleeve.

That is why General Montgomery according to an informant in the British Near Eastern Command at Cairo, wanted to know if it would not be possible for some British soldiers to dress up in German uniforms, go out into the desert, make contact with some units of the Afrika Korps, and try to ferret out the secret of Rommel's designs. A delicate job, one that required more than ordinary circumspection, tact, and courage.

If their identity was to be discovered by the Nazis, such men could not expect to be treated as mere prisoners of war. They were spies pure and simple and would inexorably meet the spy's swift death. And then: where find in the British army a group of men who could speak German so fluently that they would not be detected at the first contact with real Germans? Maybe there was a former Heidelberg student in the ranks of the Eighth Army, but of what value would one individual be on a mission like that? He would almost immediately draw attention to himself and come under suspicion, even if he should manage to enter the German lines as a free man. His usefulness would be at an end immediately.

Brigadier-General Kisch, in charge of Montgomery's supply, hearing of the project and of the search for German-speaking soldiers, told the Commander in Chief at dinner one night that he knew where to get volunteers for the job. "There are plenty of German-Jewish refugees amongst my engineers," he said, "and plenty more in the Buffs. Why don't you try them?"

Kisch's proposal was adopted. About forty German-Jewish youths from Palestine serving with the Eighth Army were examined as to fitness, perfection of language, and courage. In the end twenty were selected to go out, wander about in some section of the Western Desert ostensibly to repair barbed-wire entanglements in front of the German position and thus try to make contact with parties of Rommel's men similarly engaged, in some part of no man's land, during the lull of battle. The twenty were accordingly dressed in German engineers' uniforms, given the identity papers of fallen or captured Germans, and told to familiarize themselves with their new names. They were told to practice German military colloquialisms before setting out on their extra perilous and important venture.

Ten men and an officer disappeared into the desert and were never heard of again . . . . They must have been discovered and their real identity established almost as soon as they made contact with the Afrika Korps. The other party's fate is definitely known, for one of the ten, the officer in charge, returned. He was brought into Montgomery's presence after his harrowing experience and told the Commander in Chief: "On the fifth day out, in the neighborhood of Lahuntay we noticed a party of German engineers halting their trucks by the side of a salt depression and proceeding to lay down some road mines in a stretch of hard sand between the saline deposits and a gray mass of schist rocks that gradually rise to the west into hill-sized outcroppings of quartz. It was the only passable way for trucks and tanks for miles around, and the mines the Germans were burying were of course designed to take care of our supply columns. We worked our way unseen through the rocks and started to work on the other end of the passage about three miles away from the German party, pretending of course that we had not noticed them at all.

"An hour or so may have gone by when the Germans, having become aware of us and having spotted our German uniforms, trucks, and equipment, sent over one of their cars. We greeted them nonchalantly. But the officer asked what we were doing in that place. We told him that we had the same job as he—laying down mines—and that we had been sent out the day before by the Fifth Engineering Division of General Buchhalter's army. We gave the names of our officers, showed him our orders, and apparently satisfied him. We continued our work, planting duds. In the evening the officer of the German working party came over once more and asked us whether we intended to stay or return with him. We said we had finished our task, but that our regiment's camp was so far off that we had perhaps better stay and sleep on the spot, making the return journey in the morning, to get more mines.

"He suggested we come with him and spend the evening at his regiment's encampment, which was but fifty or sixty miles distant.

"This we accepted. We spent that night inside the German lines. We ate our meal and sat around with the men talking and smoking till eleven. One of us said that from Lahuntay as far as the eye can reach there wasn't the faintest sign of the English and that it wouldn't surprise him if the road to Alexandria were wide open. This brought on a flood of comments, of soldiers' talk, of which the general tenor was that in another ten days the advance would be under way. We made out that as soon as the last reinforcements, which had already landed at Benghazi, had joined the main army, the great assault would be launched. We asked some casual questions, giving them to understand that we knew all about the date of attack and even said that we knew to what position we would move ourselves on the appointed day. This brought replies that made it seem certain that the enemy's chief effort will not be in the coastal region but on our left flank.

They were all, it seems, going to concentrate there . . . at least the men of the units with whom we spoke.

"The following morning we drove off after breakfast and towards noon did another stretch of work. This time a British Spitfire soared overhead, swerved around a couple of times, and gave us a good strafing. We joined a work party returning to camp and had the same experience as the night before.

"On the second morning as we stood around our only truck ready to move off with the intention of regaining our own lines that day or the next, we noticed that something was amiss. We saw a military policeman on the other side of our truck examine the hood closely, lift it up, and remove the magneto. The man then walked off.

" 'Something's up,' I said to my men. 'Stand by the truck and get your guns ready. If we are trapped we'll give them a lesson. There's no use surrendering. We are spies and Jews at that. They'll give us the usual torture . . . .'

"There was no chance of making a getaway. Our truck stood in the general parking grounds of the division and was on all sides surrounded by other vehicles about to move off, hundreds of mechanics busy all around.

"I said to my men that I would go to the quartermaster's stores and see if I could get a new magneto.

"I had scarcely gone a hundred paces when the sound of firing struck my ears. I ran back, fearing that my men were in trouble. They were. They were standing in a bunch against our truck, their tommy guns in their hands, firing into a German military police detail of about fifty men who were advancing on them in a wide semicircle. From all over the camp men were running in the direction of the firing. Two of my men had clambered aboard the truck and were tossing out hand grenades with deadly effect. Germans were toppling over like nine pins. The others were firing their tommy guns point-blank into the mass of soldiers milling about. The affray did not last more than five minutes. My men's amunition gave out, and they were massacred. But around them lay a pile of dead and wounded ten times bigger than our party. I walked out of the camp as casually as I could. On the way a German major who was walking back to his hut remarked: 'That was a helluva stunt. Those fellows were Jews from Palestine. One of our policemen, a former German colonist from Sarona, spotted them this morning as they were eating their breakfast . . . .' "

One of the Jewish colonies in Palestine, Hanita, in Galilee, against the establishment of which the Palestine administration raised heaven and earth a few years back, played an important role in the war against the Vichy regime in Syria.

It was in 1938 that the Jewish National Fund acquired several hundred dunams of land in the north western corner of Upper Galilee near the borders of Lebanon and notified the British administration of its intention to establish an agricultural colony on the newly purchased territory. A beginning was to be made with the redemption of the Galilean province of the Holy Land, once the most densely settled and most flourishing agricultural area of Palestine, now for the most part a howling wilderness of rock and treeless solitude.

The Palestine administration at once vetoed the project. The High Commissioner, Sir Harold Mac Michael, based his objection on the argument that colonists taking up residence in so remote and isolated a district would present a constant temptation to Bedouin raiders both from across and inside the borders of Palestine. The nearest Jewish habitations were in eastern Galilee, too far for their settlers to come to the aid of an establishment in western Galilee in the event of danger. If the Jews, the High Commissioner intimated, instead of starting in the extreme northern wilds of Galilee, would establish colonies in the south of that province and then gradually push northward, establishing colonies chainwise or rather like steppingstones in the direction of the frontier, something might perhaps be said for the reclamation of Galilee. But to establish the first settlement at the extreme limit of Palestinian territory, in a godforsaken sort of no man's land, was a too hazardous enterprise for which he, Sir Harold, would not assume responsibility.

The directors of the Jewish Agency replied that they would be glad enough to establish an entire chain of colonies but that the administration's land-buying regulations had so far precluded the purchase of sites that might serve as steppingstones on the road to the north. They must therefore start where they could—that is, on the spot which had just recently become the property of the Jewish people. Sir Harold proved adamant. His interpretation of the mandate which charges Britain with facilitating "the close settlement of Jews on the land" works out in practice in placing, by order of the government of Great Britain, of course, as many obstacles in the way of the purchase of land by Jews as possible and after that, if the Jews still succeed in getting hold of a plot of barren, rocky, desert land on which no human being in his right senses would live, in discouraging them from settling on it.

Only, the Jews would not take no for an answer. They could not abide by the High Commissioner's decision. In withholding his official fiat, the High Commissioner may well have been carrying out his duty in that he acted in the spirit of those restrictive measures designed against and imposed upon Jewish Palestine by a narrow-minded, anti-Jewish bureaucracy in the sole interests not of the British Empire, but of a handful of feudalistic Arab landowners. The Palestinian Jews, on the other hand, could not do otherwise than what they did. They insisted again and again that the High Commissioner's decision be revoked and that the colony be opened up—that particular colony and others, always more colonies and settlements by hook or by crook. For the Palestinian Jews feel behind their backs the ever-growing anguish and desperate pressure of the homeless and hopeless Jewish masses in Europe still seeking a way out of what had become to millions of them a gruesome deathtrap or a living hell after Hitler's advent to power . . . .

After months of wearisome palaver, pleading and insisting on the one side, haggling and quibbling on the other, with references to the Colonial Office in London going to and fro, the High Commissioner finally, reluctantly gave in. The Jews were permitted to establish that colony on their own land in their own country. They could go out there to that desolate spot in Galilee if they wanted to, but they must not blame the administration if disaster should overtake them on the pioneer trail.

"Mi yivne ha-Galil? Who will build Galilee?" the young people sang that night all over Palestine, when the government's decision became known.

"El yivne ha-Galil! God will build Galilee!" came the answering chorus.

The tract of four thousand dunams had been thoroughly explored and surveyed in the meantime. It would provide a living for eighty families, or five hundred souls, if they could engage in mixed farming: sheep raising and poultry breeding, with tobacco the chief crop. All this had been settled by the agronomical experts who had examined the land. One third of the area was to be used for pasturage, and one of the first tasks of the settlers would be the planting of a forest of eucalyptus trees. For deforestation and consequent soil erosion constitute one of the worst blights of the Holy Land. The candidates to take up the work were in readiness, too. The occupation group consisted of ninety young people, eighty men and ten women. They were to go out before the bulk of the settlers and make the place fit for habitation. The pioneers had been carefully selected from many localities with reference to their fitness and courage for occupying a new tract in a frontier region where only recently fierce battles had raged between government forces and Arab bands.

The occupation took place in what has become of

late years the usual form for establishing a new settlement in Palestine: the colony was completed in all essentials between sunrise and sunset in one single day. All preliminary preparations were made in the workers' quarter in Emek Zebulon, at the foot of Mount Carmel. The caravan of trucks was on the way while the moon still hung over the dark waters of the Mediterranean. Thirty-seven lorries loaded with tents, planks, mattresses, cots, length of iron pipes, provisions, and water rumbled off into the future. The orders were that they must stay closely together, that there was to be no singing on the road, and that no one change from one truck to another. At the head of the procession rode a party of ghafirs, or supernumerary constables, themselves Jewish pioneers. Behind them, in motorcars, were four hundred laborers, who were to return in the evening after the colony was established At the tail end, behind the trucks, trotted a contingent of donkeys needed to carry loads up the hill. A second group of ghafirs brought up the rear.

The sun came up as the party arrived on the chosen spot. Immediately the workers scattered to tasks previously assigned to them. Some began to mark out the by-road which was to connect the settlement with the Haifa-Beyrouth highway running along the Mediterranean shore, less than a mile distant. Two springs had been discovered by the surveyors. Their waters were now brought to the camp by means of pipes. Tents were pitched, weeds uprooted, stones pried loose for a barricade behind a barbed-wire fence which was going up in the meantime. The high wooden watchtower was set in place with a giant searchlight. Some of the ghafirs stood on guard while the rest hammered and shoveled and got to work, although their rifles remained near by for handy reference, if need be . . . .

At midday a recess is called. The food is quickly eaten, for there is still much to be done before nightfall. But a few minutes remain before the back-to-work signal will be given, and someone starts to sing "*Alinu hartza,* we have gone up to the land!" Instantly the Hora dance circles form, widen, expand, whirl like wheels within wheels, faster and faster. But not for long . . . .

"*La avoda,* back to work!" Hammers pound, saws buzz and rip, picks clang their way into strong ground. All hands are working at top speed. There is not a minute to lose. The settlement must be completed before nightfall, and sun has now passed the meridian. A fresh breeze comes up from the Mediterranean. It must be near five o'clock, the hour when the old-world sea always stirs mysteriously, no matter how calm the weather.

But then there is an interruption. A delegation has come to the newest settlement in western Galilee from the oldest in eastern Galilee, bringing a gift of a Sefer Torah, a scroll of the Law. All work comes to a standstill. Quickly a tent is cleared out and converted into a synagogue. Six young men advance and in turn kiss the scroll which is contained in a cylindrical velvet-covered, tubelike box with a ring of little silver bells fastened to the top of its axis. The whole company forms a procession to take the Word of God to its new home.

All at once the man bearing the scroll breaks into singing: "*El bene, bene betcha bekarov,* build thy house, O God, build it speedily!" The congregation takes up the prayerful chant, and they dance as long ago their king, the "man after God's own heart," once danced when the Ark of the Covenant was brought to Jerusalem.

"I was glad," shouts one man, with arms outstretched and dancing about, "I was glad when they said unto me: let us go into the house of the Lord!" All the colonists repeat the words of the One Hundred and Twenty-second Psalm: "Pray for the peace of Jerusalem: they shall prosper that love thee . . . . "

Now the sun is going down swiftly. The visitors and helpers from Haifa are leaving. Most of the ghafirs are piling into their cars. That night five troopers will remain behind with the ninety young men and women in the wilderness. Darkness gathers quickly in the Holy Land. The colonists are in their new home, for better or for worse . . . .

For worse, it seems! At midnight the colony is attacked by a large guerrilla band. The prowlers are spotted by the revolving searchlight when they are only two hundred yards distant and in the act of dividing themselves into three columns for the assault on the colony. The watchmen shout the alarm. Instantly the night is filled with the warcry of the Bedouin. It comes from all sides. The attacking force must be a thousand strong. The searchlight beats down on the white-cloaked figures now running at top speed. They are coming from three sides and firing their rifles. But the Jews have manned the barricades. They are holding their fire. Suddenly the searchlight goes out. The attackers are too near now to let the light burn: it would give away the position of the defenders. The Bedouin blaze away aimlessly. But now there are short spurts of flame from the rocks of the barricade: the Jews have held their fire till their targets are unmistakably clear. Two, three, four salvos—and then stillness.

The searchlight goes on and slowly turns its beam on the surrounding country: the Bedouin are streaming back. The firing has stopped altogether. Nothing is heard but the moaning and whimpering of the wounded. The beam points fiercely at a fallen Bedouin who is trying to raise himself on one hand while keeping the other convulsively clamped over his abdomen. When he becomes aware that the light keeps him in focus, he turns his face in the direction of the camp and cries out in a whimpering voice: "Don't kill me, *baba!* Don't kill me, I am your brother!" The light moves abruptly away from him. It sweeps, without stopping, over what appear mere bundles of white rags and then peers into the distance, lighting up the bluish hills opposite.

The Bedouin are assembling for a new assault. The three columns are now merging into one solid phalanx. There is to be a mass attack this time. It is to come from the north side of the colony, the side where the barricade has not been completed. The first attackers have discovered the colony's weak spot. They now come running forward en masse, filling the night with their unearthly screeching. The searchlight beats down on them. When they are four hundred yards away they drop to the ground and begin to disperse, crawling on hands and knees, taking advantage of every clump of shrubbery and of every boulder. Whenever the searchlight is turned in a southerly direction for a moment, to see that no enemies are sneaking up from the rear, the attackers in front, taking advantage of the temporary darkness, dash forward a few yards. The whole plain is filled with creeping, crawling figures.

At a shout from one of the leaders they get to their feet and rush for the opening in the fence. The first men are fifty yards away when once more they drop to the ground. Now they fire their guns. The searchlight is bespattered with bullets. In a flash the Bedouin are up and running again. They are only twenty yards off. Their faces can be clearly distinguished. Many carry a knife between their teeth. That will be for throat-cutting and mutilation when they get within the colony. But now with a simultaneous crash the colonists' rifles spit fire. The foremost ranks of the attackers go down with a terrifying shout, the second and third waves push on over the bodies of the fallen. Another salvo from the colony, as deadly as the first. It is immediately followed by a third and a fourth. The Jewish fire is sustained. Some Bedouin are right on top of the barbed-wire fence. One of them shouts to his followers: "We are in!" as he flourishes a long knife. At the same moment he grabs his throat and falls. A Jewish girl has shot him through the mouth. The bodies of three or four of his immediate followers fall on top of him, a writhing cluster of wounded and dead.

The others still out in the field begin to waver. They fire their rifles and run back; first, individuals only, then the entire band turns and flees. The Jewish fire keeps up relentlessly. Every shot finds its mark among the retreating guerrillas. Dozens of white-cloaked figures slump forward. They do not rise again. Soon the night is still, except for the cry of wounded men.

Once more, just before dawn, the horde returns. But again it is bloodily repulsed. When a British police patrol arrives at seven o'clock, the colonists are watering the donkeys and rigging up the rest of the barricade. Others are busy putting the roof over the tool shed.

"What happened here last night?" asks the officer commanding the patrol.

"We were attacked!" replies one of the colonists.

"That's rather obvious," says the officer. "The whole plain is covered with dead men. We surprised the last Bedouin carrying off their wounded . . . . Have you any casualties?"

The colonist points in the direction of the tent where the scroll of the Law was placed the previous afternoon. "Eight dead, one wounded; he is dying too!"

"Well, this is no place for a colony anyway," says the officer shaking his head. "What will you do if they come back tonight?"

"We will be ready for them!" says the Jew quietly, as he drives a stake into the ground.

The officer looks into the tent where the dead lie and takes out a notebook. He writes something down in it, walks out, swings back into the saddle, and rides off with his men.

Thus ends the first day in the life of Hanita, the name by which the colony is to be known. Hanita means *spearhead,* a pointing arrow.

The Bedouin do not come back the next day or the night after that. Before a week is over the colonists have built a blockhouse from the boulders gathered from the land. It stands in the middle of the colony like a citadel. It houses the administrative office and the hospital, but it can also hold the entire colony's population of four hundred souls if they are attacked again. But the rest of the year 1938 goes by, and 1939 too, and no attack comes. The schoolhouse has been built; stables, the little synagogue, and the storage shed also rise. The first crop has been harvested. Hanita grows: a park has been laid out, and a theatrical company has been giving performances for a week. The trees in the eucalyptus forest come to the height of a man's chest. Another colony has just been started at a distance of three miles. The year 1940 has just dawned . . . .

A British patrol rides up one day. The officer dismounts and enters the offices in the citadel. He wants to speak to the colonists. The men gather around him. He speaks bluntly and to the point. "Have you men sufficient knowledge of the trails leading into the Lebanon country to guide the Australian troops? There is going to be an attack on Syria. The Vichy French are letting German 'tourists' into the territory of the Syrian Republic . . . . They are laying out airdromes for the Luftwaffe . . . . Dozens of German transport planes are coming in daily across the border with men and equipment . . . . Britain has not declared war on Vichy yet. But Britain cannot tolerate the attitude of the Vichyites much longer. She cannot permit a German army near the Mosul oil fields and the Suez Canal. Australian troops are moving up from the south now. There is going to be an invasion of Syria. The High Commissioner wants to know whether the men of

Hanita are willing to serve as guides . . . . "

The High Commissioner! Sir Harold has apparently forgotten the long squabble he had over the establishment of the colony. Hanita is going to be a base for the Australians attacking Syria . . . .

Two days before British forces, in collaboration with the Free French, invaded Syria, fifty young Jews from the colony of Hanita struck out across country, entered the territory of the Lebanese Republic, avoiding all Vichyite frontier posts and patrols. After a march of seventeen hours, they halted in sight of Fort Gouraud. This stronghold, which is built on the site of an ancient Crusaders' castle in the foothills of the Lebanon range, dominates for a space of ten miles the hard, sandy highway running between Haifa and Acre, in Palestine, and Tyre, Sidon, and Beyrouth, in Syria. The road is flanked on the west by the Mediterranean whose waters it skirts, and on the east by sand dunes and the steep rocky crags of the Lebanon chain. Within range of Fort Gouraud's guns lie three bridges, two to the south and one to the north. These bridges, which form part of the coastal highway, link the steep banks of three swift mountain torrents which, decending from the Lebanon, throw themselves over a series of cataracts and rapids into the sea.

The Jews of Hanita had been asked by the British military command, because of their familiarity with the wild and unchartered Galilean-Lebanese border country, to take up a position in the neighborhood of Fort Gouraud and, if possible, to prevent the three bridges from being destroyed by the Vichy French. For it was along the Acre-Beyrouth highway and its three bridges that the invasion's Australian spearhead was to advance into Syria. The destruction of the bridges would naturally hold up the invasion and might in fact well turn it into disaster, as it would mean leaving baggage, trucks, artillery, and tanks behind. Fort Gouraud, moreover, is so placed that a column of troops marching north along the Beyrouth highway can be taken under the fire of its guns long before any effective reply can be given.

The task allotted to the Jews was therefore not only difficult but also extremely hazardous and, one may well add, foolhardy in the extreme. Everything depended on seizing the bridges by surprise and holding them, under the fire of Fort Gouraud's guns, until the Australians should come up in force along the highway. But the British command knew well what it was doing when it chose the Jews from Hanita for this risky undertaking. The leader of the Jewish francs-tireurs, Moshe Dayan, and ten of his companions had just been liberated from prison in Acre, where they were serving life sentences for . . . having explored the Lebanese border region and for having engaged in scouting activities in connection with the defense of their colony. Nearly all the Jewish scouts who served the invading British and French forces in the invasion of Syria—that is, the guides of the seven spearheads—were Jews released from jail where they were serving long terms for having prepared themselves for the precise task they were now called upon to carry out. They called to mind that simple and heroic man Raziel,* the com-

**That Day Alone,* Garden City Press, New York, 1943.

mander of the secret Jewish defense force, who was taken from jail by General Wavell's order to go into Iraq at the time of the Rashid Ali revolt against Britain and keep the Arab allies of Hitler from destroying the invaluable oil refineries of Mosul....

Having come in sight of Fort Gouraud, Moshe Dayan and his men first explored the neighborhood. They noticed that although the Vichyites guarded the bridges but weakly, on the ramparts of the fort, a thousand yards off, there was a row of machine guns pointing their nozzles in the direction of the bridges, which would make it most difficult if not to seize them, at least to hold them for any length of time. They therefore decided to take the fort itself and thus make themselves master of the position that controlled the entire district. This was easier said than done. There was a garrison of three hundred tough Somali troops in Fort Gouraud and although many of these were periodically absent on patrolling and reconnoitering duty in the neighborhood and on guard at certain strategic points, there remained a good hundred men within the walls of the fort within easy call of the ramparts with their artillery and machine guns. All the arms the Jews had were ten tommy guns.

Nevertheless, they would take the chance. There seemed no alternative. The Australian column, they knew, had left Haifa and would presently move headlong into the range of Fort Gouraud's guns, for the highway was exposed to the view of the Vichyite sentries and lookouts for a distance of five miles on either side. The Somalis could make all the preparations to have the Australians walk into a first-class deadly ambush.

This must not be! Moshe Dayan and his men therefore, crawled up to the fort's gate at the crack of dawn, knocked the sentries over, overpowered the guard, and raced for the ramparts. There they seized the machine guns, swung them around, and pointed them in the direction of the small parade ground between the barrack buildings. As they busied themselves on the ramparts a few shots were fired at them from the windows of the brick barracks, but when the Jews let the machine guns talk back and sent a few squirts of bullets into the building the Somalis came out with upraised hands and surrendered.

The Vichyites were then ordered to pile up their rifles and small arms in the square and to remain indoors. Fort Gouraud was taken. And Dayan hoisted the Union Jack as a signal to the Australians when they should come in sight that all was well. This was an error. For the Somali outposts guarding the bridges, having heard the sound of machine gun fire inside the fort and now seeing the British flag aloft over the main building, promptly set off one of the prepared charges under the southernmost bridge. With a deafening roar and clanking of steel and in a cloud of smoke, the girders collapsed. Dayan, realizing that he had been outwitted, immediately took the other two bridges under fire and thus prevented the Vichyites from crippling them in turn. He intended to keep up his fire till the Australians should arrive.

However, he had not noticed that some of the prisoners locked up in the barracks had had an opportunity to escape by a door in the cellar. These men had joined their comrades' patrols outside the fort and informed them that the enemy was a mere handful of nonprofessional soldiers and that a joint effort could not fail to expel the bold intruders. By noon all the Vichy patrols in the neighborhood had been assembled, several hundred men in all, and were making ready to take the fort back.

They began by crawling through the shrubbery and picking off the Jews on the ramparts one by one. In a short time Dayan's force was reduced by seven men, more or less seriously wounded by the Somali sharpshooters. Then the attackers reentered the fort through the same cellar door through which the prisoners had escaped a few hours earlier. They rushed upstairs into the barrack building, mounted machine guns they had brought along, and took the ramparts under fire from the rear. Dayan now was compelled to switch his own guns around in order to defend himself. Five more of his men fell. But now the Somalis were fixing bayonets, and their snipers were taking up advantageous positions in the square, on the roofs of buildings, and in the stock rooms of the arsenal. A hail of bullets whizzed about the heads of the Jews on the walls. Once, twice, a concerted bayonet attack was beaten off, but each cost Dayan ten of his men. It was obvious that he could not hold out and that sooner or later he would be overwhelmed and annihilated. In desperation he planned to make a sortie, get out of the fort, rush for the bridges, and hold them perhaps another hour, two at the most, until the Australians should finally come up. But the sustained fire from the barracks windows made the idea of leaving impossible. Once more he trained his field glasses on the highway. It was getting dark. The ammunition was running out. The Somalis were coming nearer and nearer. Only twenty-two of his men were able to stand up. Another hour, nay less, and then the end would be there. But then, yes, there six or seven miles away was a cloud of dust. "They are here!" shouted Dayan. As he spoke a bullet struck his field glasses and drove the instrument into his eye. He tore it out and the eye with it.

At the same moment the Somalis launched another attack with the bayonet and with hand grenades. This time they reached the ramparts and actually got on the walls. The Jews fought them off in a hand-to-hand encounter, with knives, with bricks loosened from the walls, with crowbars and revolvers. As they were forced into a corner, two Very lights, first a red one, then a white bulb hanging in the sky beneath its floating parachute, flew overhead. The Australians had surmised what was going on in the fort and had sent up the signals. At once the French surrendered a second time. It was high time. Dayan's men had thrown their last grenade. They were fighting with their bare hands. They had held off the enemy for seven hours and resisted ten mass attacks. In the night the bridge was repaired, and the invasion of Syria by British regulars was on its way.

That's Moshe Dayan, incidentally, the man with the black bandage over his left eye who runs the gasoline station in Colony Hanita—Colony Spearhead—these days . . . .

There was a Jewish battalion among the defenders

of Tobruk during the long siege . . . .

It was Major Richard Perach, a Jew from Talpiot, who led the battalion which turned the flank of the famous Mareth Line . . . .

Jewish suicide-task forces landed in Tobruk and helped capture the city as they had captured Bardia earlier . . . .

It was General Frederick Kisch, a Jew from Haifa, who organized and supervised the immense undertaking of Montgomery's supply in the battles of Egypt and on the 1300-mile trek through Libya and Cyrenaica right to the gates of Bizerte, where he was killed . . . .

Jewish engineers organized and manned the coastal defense and signal services of Lebanon, Palestine, Sinai, and the Red Sea coast.

Entire Jewish families in Palestine enlisted in the British service—fathers and sons, mothers and daughters—in the medical corps and the Red Cross.

The Jewish coast guard ran one hundred speed boats along the dangerous Mediterranean lines between Cyprus, Palestine, Egypt, and Libya . . . .

Two thousand-five hundred Palestinian Jews served with the Royal Air Force as bombardiers, pilots, and observers. Six thousand more Jews were in the ground crews of the Egyptian airdromes . . . .

Jewish engineers constructed Montgomery's impregnable forts at El Alamein . . . .

All the antiaircraft stations in the Holy Land were manned by Palestinian Jews . . . .

Jewish units took a leading part in the capture of Sidi Barrani, Sollum, and Fort Capuzzo and were praised by General Sir Archibald Wavell for "their courage and self-sacrifice" in those engagements.

Jewish "suicide squads," especially chosen for their toughness, daring, and mobility, fought in Eritrea and Ethiopia.

"On reaching the Mogarreh Valley the Palestinians," one dispatch to G.H.Q. reported, "had the task of covering the left flank of the advance on Keren, and the cutting off of the Italians on a ridge to the left of the main attacking force. They thus helped to take 2500 prisoners and bring to a successful conclusion an extremely tough and courageous action."

A Jew, Shmaryahu Weinstein, was the hero of the battle of Keren, in Abyssinia: he saved a whole company of South Africans by sacrificing his own life.

Thirty Jews were killed in the Amba-Algai area in Ethiopia when they, forming a suicide-task force, rushed forward to capture a hill that bristled with the machine guns of the Duke of Aosta's forces.

Jewish units—suicide-task forces—penetrated and demolished enemy fortifications night after night during the campaign in Abyssinia. Peter Fraser, Prime Minister of New Zealand, said that his country and Australia were proud of the way the Palestinian Jews stood shoulder to shoulder with New Zealanders and Australians in "the hottest engagements" in Greece, Crete and Erithrea.

Jewish engineers, the Jewish Pioneer Corps, and Jewish pilots fought in Greece. Their services and severe losses were recognized by Sir Henry Maitland Wilson, the commanding officer in Palestine.

Three thousand Jews, residents of Palestine, joined the national armies of the countries to which they owed allegiance—Czechoslovakia, Holland, France, Poland . . . .

It took the war and, above all, Marshal Rommel's grave and long-lasting threat to inter-Empire and inter-Allied lines of communication to set in strong relief the value of the settlement of half a million Jews in Palestine—that is to say, in the immediate vicinity of the most vital arteries of the British world system. Jewish Palestine contributed more effectively to the repulse of the Axis advance upon Syria, Iraq, Egypt, and the Suez Canal, to the triumph of British arms in Ethiopia, Eritrea, and Somaliland, and to the final liberation of the Libyan and Tripolitanian coastal regions than the fifty times larger populations of the Arab countries combined.

In all that Eastern world where hazardous contingencies lurked around every corner or lay carelessly strewn about as so much explosive, the Jews were the only ones who took their stand at Britain's side without hesitation or flinching. They responded to the call to the full measure of their ability. They did more: they responded before a call was made. No more than the Arabs of Egypt, Iraq, or Trans-Jordan were the Palestinian Jews, by reason of some existing tie of allegiance or loyalty, bound to espouse the British cause. They were neither British subjects nor nationals of one of the United Nations allied with Great Britain. They were merely inhabitants of a territory which is temporarily administered. Nothing could be expected of them beyond compliance with the general laws in vogue. According to international law, Palestine is a neutral country, and its inhabitants are neutrals.

It is true that the Palestinian Jews could not have acted otherwise than they did whatever their legal position, inasmuch the moral issues involved in the war determined their choice. Like the prophets of old, the Palestinian Jewish community measured its duty by the objective morality and not by its own intrinsic interests. What were they to gain in allowing the best of the limited Jewish man power in Palestine to throw itself into battle when there did not exist a guarantee, or even the prospect of a guarantee, that Britain would permit a replacement of the inevitable losses by a greater volume of immigration than that stipulated by the White Paper of 1939. The Palestinian Jews were under no illusion, when they placed their lives, their wealth, and their honor at the disposal of Great Britain, that the gesture would be graciously returned with at least a promise of the discontinuation of the destructive policy pursued by the British Colonial Office in the building of the Jewish national home. In their case truly the only recompense was blood, sweat, and tears. Even hope, which is the most puissant incentive of the other peoples engaged in the struggle for liberation, was denied them.

It may be objected, of course, that the contribution of so small a country as Palestine with a population numerically insignificant when measured against the fantastic number of human beings involved in the war cannot have been of decisive import. It may also be argued that it is too much to believe that the Jews were motivated entirely by a sentiment of altruism when they sprang to the Empire's defense. But to this the reply must be made that when a sensitive pair of scales is evenly balanced, the smallest straw will upset the equilibrium. Jewish Palestine was that straw. The forces engaged on either side in the Near East were not at all commensurate with the gigantic scope of the world struggle. Rommel, who represented the chief menace, never had more than 200,000 men under his command in Africa. Until Montgomery took over from Auchenleck, the British, Mr. Churchill has declared, had at no time more than 45,000 men on the African front. Of these 45,000 a quarter were Jews.

Not until the joint Anglo-American landing in Morocco and Algeria did the British have more men and material in the North African-Near Eastern theater of war than their German adversary. But by this time the scene of battle had shifted thirteen hundred miles away from the Egyptian border. Tunisia is not in the Near East. The battle for Egypt and the Suez Canal was won by Bernard Montgomery before American troops set foot on Africa's shore. It was the British Eighth Army that turned the tide. In that army and supporting that army were thirty thousand Jewish volunteers and two hundred thousand Palestinian industrial workers and farmers. In the circumstances that was not a negligible force.

As to the basic motive of the Palestinian Jews in throwing their lot with the British Empire: the deed itself is not supremely significant; decisive are the intention and the soul of the doer. What counts is not what is done, but how it is done. The Jews of Palestine collaborated instinctively, spontaneously, and wholeheartedly. Renan once said: "That which causes men to form a people is not only the recollection of great things they have done together, but the longing and the will to do new things."

Intuitively, the Jews of Palestine responded to a moral challenge, to a call, to the ache of national community and common destiny a thousand years old, thwarted and vitiated ever so long in the Diaspora and now suddenly again become vivid, clear, and alive when confronted by a common task. Without words or phrases, but by the spontaneity of life, they vindicated the fundamental tenet of the philosophy of Zionism that through being rooted in their own land they had become again what they always were, but now in sight of all the world: a people, a nation, Israel reborn . . . .

In the fateful hour when Britain's fate hung in the balance and when Rommel boasted that as far as he was concerned it was all over but the shouting, the little land of Palestine placed at the disposal of the British Empire and its armies in the Near East an industrial apparatus of seven thousand factories, large and small.

It may sound fantastic, but it must be said, for it is the truth: nobody, neither the British High Command nor Marshal Rommel, had counted on Palestine to play the role it did in that battle which decided—future chroniclers will declare—the world's destiny. Of course, everybody knew that Palestine existed and what its resources were. But it was not realized what Palestine meant: a powerful instrument of victory in the hands of that Jewish community, which threw itself into the struggle with a wholeheartedness and zeal that multiplied the country's industrial effectiveness a hundredfold. It was the spirit that counted.

The pioneers of Palestine showed the British government, which had treated them most ungenerously, that they had not only exploited the poor natural resources of the Holy Land to the utmost, but that they had succeeded in creating new resources which had not been available twenty-five years earlier when modern Zionism launched its program of redeeming the land of Israel. Sandy wastes had been planted and cleverly cultivated. Desert solitude and grim rocks had been transformed into vineyards and orange groves. Swamps and malaria-infested regions, twenty-five years ago pestilential deathtraps for the Turkish and British armies engaged in the battles of Gaza and Megiddo, had been turned into healthy and productive areas. Wells had been dug so that no intricate and endless system of pipes had to be constructed across the Sinaian desert as when Allenby's campaign was held up by the lack of water. The rivers Jordan and Yarmuk had been electrified, the latent wealth of the Dead Sea made productive, and the godforsaken wilderness transformed into a garden fit for human habitation. As if guided by a providential foresight, the Jews had constructed a system of roads, as good as and better than Mussolini's famous *autostradi* in Libya, branching out and radiating in a hundred directions and leading up, in several critical instances, to the very borders of adjoining countries that were directly and most gravely menaced by the enemy on different occasions in the course of the present war.

The world will learn some day—and should remember at the peace conference—how certain Arab chieftains in Iraq, Egypt, and Saudi Arabia, who proclaimed themselves the loyal allies of Britain and of the United States after Rommel was beaten at the gates of Alexandria, comported themselves when Montgomery himself said privately that only a miracle could save him . . . .

Jewish Palestine was part of that miracle. Jewish Palestine was one of the imponderables that turned the tide against Hitler at the moment when he and almost the whole world least expected it.

Besides the thirty odd thousand Jews who were taken into the Eighth British Army and into the Palestinian home-guard forces, Jewish industries large and small alike and Jewish agriculture in Palestine supplied the British armies with bandages, pharmaceutical articles, ether, sulfanilamide, benzoic acid, nicotinic acid, vitamin B complex, ascorbic acid (vitamin C), insulin, alkaloids, microscopic stains; further: precision instruments (many of the experts of the Carl Zeiss works of Jena, having transferred to the Holy Land a few years previously), tobacco, 135,000 pairs of boots per month, fruit, vegetables, wheat, wine, X-ray apparatus, 25,000 tons of cement per month, soap, chocolate, spare parts for automobiles and trucks, sandbags, timber, tents, linen uniforms, and buses taken from the Palestinian transportation system.

The Concrete Shipbuilding Company, organized by a refugee shipwright from Serbia, launched a fleet of twenty fishing trawlers, of 110 tons each, equipped with Diesel engines also manufactured in Palestine. The concrete manufactured by the refugee builder is seven hundred per cent more watertight than ordinary concrete and can withstand a pressure of five hundred kilograms per square centimeter. These vessels are built in one quarter the time required for steel vessels, and the construction cost is forty per cent less. In the beginning of 1943 the company was going in for mass production of concrete barges, tankers, launches, and cargo vessels.

During the war new items and new techniques of manufacture were introduced in Palestine in turning out hydrogenated vegetable oil for the manufacture of margarine, glycerin, yeast, starch, glucose, waterproofing material, castor-oil processing for lubrication of airplane motors, and rubber hose. Ethylene gas and ethylene chlorohydrin were turned out for ripening and preserving certain crops. A new distillery was opened during the war to manufacture alcohol from carob expulp. The gum manufactured from carob seed is exported to the United States.

In 1942 nearly two million cases of waste citrus fruits were utilized for the production of pectin, alcohol, essential oil, citric acid, jams, jellies, marmalade, and dried cattle fodder.

With that Palestine sent out to the Eighth Army: laboratory equipment, electrical appliances, meat, especially mutton; photographic material, water pipes, stationery, typewriters, burr drills, steel helmets, three million dollars' worth of processed diamonds for cutting tools, beer from Palestine's two new breweries, matches from Emek Zebulon, glassware and crockery, bedding, mattresses, maps, sheet metal, parachutes, disinfectants, oil burners, armored cars, cranes, five hundred ambulances, wire, shovels, water pails, cutlery, hammers, saws, nails, screws, saddles, camels, mules, horses, and railway equipment and steel from the Vulcan Foundry and Forge Works at Emek Zebulon.

All that material and equipment, as necessary to an army as bullets and bread, did not need to be transported through submarine-infested oceans. It was there, on the spot, at the moment it was needed most.

Palestinian Jews lugged the oil and gasoline from Haifa and Mosul across the Syrian and the Sinaian deserts to Montgomery's mechanized forces.

Palestine furnished to the British armies in Libya, Eritrea, Ethiopia, and Somaliland, thousands of doctors, nurses, and dentists. It placed the great Hadassah Medical Center in Jerusalem at the disposal of the Empire.

The Meteorological Department of the Hebrew University prepared weather data for the British and Allied air forces operating in the whole Near and Middle East, in the Caucasus, the Sudan, Ethiopia, and Eritrea. The Hebrew University's Department of Parisitology conducted courses in war surgery and tropical medicine for the Australian Expeditionary Force. It supplied antitetanus serum and typhus serum to the Army Medical Corps and shipped 70,000 phials to the Polish army in Russia and as many phials of typhus serum to the Red Army. The Department of Oriental Studies furnished the intelligence service with interpreters familiar with all the languages and dialects of the neighborhood, including Somali, Amharic, and Galla for Ethiopia, Kurdish, Chaldean, Turkish, and Armenian for northern Iraq, and Coptic and Berber for northern Africa.

Palestine supplied the Turkish army with half a million pairs of boots and all the pharmaceutical articles and precision instruments Ankara required. To the Soviet Union went large shipments of bromine compounds, indispensable in the manufacture of explosives, a gift from the Palestine Potash Company, which exploits the resources of the Dead Sea.

Palestinian Jewry spent the equivalent of about ten million dollars to mobilize fresh resources for agricultural and industrial developments in the first two years of the war. It started twenty-five new agricultural settlements in the same years. It completed seventeen hundred new electrical power installations in the southern districts of the country alone.

What Lord Strabolgi called "Palestine's prodigious contribution to Britain's cause," did not materialize overnight, although its effect was felt in a sudden, startling manner in an especially critical hour of history. It was the result and culmination of twenty-five years of patient planning, building and sacrifice on the part of the Histadrut or Palestine Federation of Pioneers Workers & Cooperatives. Primarily to the Histadrut, which controls and guides and co-ordinates fully eighty per cent of all social, agricultural, industrial and cultural efforts in the Jewish Home Land, must go the credit of having performed the marvel of having transformed a backward and barren area into a flourishing and modern industrial and agricultural community, of having built a house for Israel, but also of having provided Great Britain and the United Nations with that advance post of democracy which served as the strongest basis of defense in the Middle East in the struggle against the Nazi designs of world domination.

For so remarkable and telling a show of devotion rendered to the British Empire, the least that could be expected is a public recognition of Jewish valour and an acknowledgement of Palestinian Jewry's share in that momentous victory which frustrated Germany's plans to break through to the sources of raw materials.

Wendell Willkie said that the measure of freedom to be gained by the peoples of the Near East in the postwar era—and the Jews are one of these peoples—will depend on what contribution they made in the actual waging of the war. Willkie said that; not Churchill or Eden or Lord Cranborne, the Colonial Secretary. These men were silent. The British press was silent. Not a single British, or American newspaper gave the merest inkling of the material and strategic importance of Palestine and the role it played in this theatre.

Never once was Jewish heroism mentioned. Not a word leaked out of that numerically small Jewish community standing there like a rock in the smoldering hostile Arab world at the back of the British army. As little as possible was said of Jewish soldiers standing side by side with Englishmen, Australians, and South Africans and facing the fearful odds presented by Rommel's overwhelming superiority. And yet the Near East swarmed with foreign correspondents from the beginning of the campaign to the end.

Palestine Jewry's effort in the war must be considered the best-guarded military secret of all!

CHAPTER 5

IMPERIALISM'S REWARD

BUT why that ungenerous, niggardly attitude, that officially inspired mutism? Britain, or its ambassador in Washington, Lord Halifax, who knows what is going on in the United States, could with one word have silenced the whispering campaign set in motion by Dr. Goebbels according to whom "the Jews are quite capable of starting wars but leave the fighting and all honest work to others." The elaborately organized British Information Service in the United States could have pointed to Palestine, a country under British mandate, whose Jews did fight, where out of a Jewish population of half a million, one hundred and thirty-seven thousand registered for service the moment war broke out and whose Jewish inhabitants fought to the full extent they were permitted to fight.

What did Britain seek to hide? Wasn't Hitler, who has slaughtered a larger number of Jews than the British Empire has suffered in casualties, to know that Jews are men who will fight back when they have arms in their hands?

Weren't the Arabs to know that even if Britain could not count on them, she could rely on the Jews, till the bitter end if need be! What somber and sinister game is being played behind that shield of indifference and silence on the subject of Palestinian Jewry's contribution in the war against the Axis?

That legitimate information on and recognition of Palestine's achievements and Palestinian Jewry's role in the war was designedly withheld by the British authorities was clearly shown in May, 1943, when the delegates to the Conference on Refugees in Bermuda were instructed to leave Palestine out of consideration in their "exploratory" canvass of the four corners of the earth for a prospective haven of refuge for the Jewish masses under the Nazi terror in Europe.

At that conference the British delegates sought to persuade their American colleagues to recognize the validity of the British government's White Paper, which bars Jewish immigration into the Holy Land, beginning with the spring of 1944.

Jewish Palestine was therefore not to be mentioned in dispatches and the support that Palestine Jewry gave unstintedly to the British Empire was to be ignored because the Colonial Office wants to divert attention from Palestine Jewry's role in order to forestall any possible Jewish demands. These might logically want to scrap the White Paper and admit, as soon as possible, the weary wanderers of Israel into that land where the economic, social and religious framework for their reception and integration has been built, is functioning, and is ready to receive them.

If the British Colonial Office is to have its way, the Jewish people are to be confronted with an unassailable and unchangeable *fait accompli* in the spring of 1944: the remnants of the martyred Jewish masses in Europe who have nothing to look forward to but the last great exodus to their Promised Land are going to find the gates to freedom and life shut in their faces.

And it is not Hitler, I am deeply sorry to say, but the government of Great Britain that is carefully and cynically planning to perpetrate this almost unimaginable piece of cruelty. Britain is not waiting for some peace conference or some new conference on refugees to discuss and debate the problem of the Jews of Europe, where they are to go or what is to become of them after the war.

The British government is trying to settle that problem now, while the war is still going on, in an arbitrary manner, when there is no opportunity to convoke Danish, Norwegian, and Dutch members of the Permanent Mandates Commission and when The Hague Court is not in session. They are trying to settle that problem without consulting either the internationally recognized Jewish Agency for Palestine or the sovereign states who committed themselves, by ratification of the mandate, to participate in an attempt to solve the problem of Jewish national homelessness by providing Israel with a national home in the Land of Israel.

No sooner had the threat to Egypt and Suez been successfully parried by the brilliant victories of the Eighth Army than the British government began laying plans to carry out a vast program of reconstruction in the Near and Middle East. With this singular display of haste it betrayed the insincerity of the answer generally given to those who insist that planning for reconstruction in the postwar era should begin at once

—that is to say, without waiting for the end of hostilities.

The usual answer to these representations is that a discussion of the shape of things to come is inexpedient and premature while actual warfare continues. Only when the smoke and dust of battle have subsided, when emotions and passions have been assuaged in a more or less lengthy period of cooling off, when reason may be expected to have replaced blind rage, when calm has returned and the turmoil of daily catastrophic events can no longer distract the studious deliberations of commissions, conferences and parliaments, has the time come to think of the future.

Public discussions at a moment when the enemy is still capable of an enormous show of force would be detrimental, it is objected, to the singlemindedness of the citizenry. Deliberations on what form, character, and shape the world of tomorrow will or should assume cannot be profitably undertaken until the utter defeat of the Axis has been accomplished. No energies should be wasted, no deflections allowed to sidetrack the great goal in view.

It is a good argument, for it appeals to the patriotic sense of the vast majority of Englishmen and Americans. But it is a trick and a cruel hoax that is being perpetrated on the peoples of the world. For while discussion is thus silenced or allowed to course into channels of innocuity and harmless faddism, the governments are doing the very opposite of what they preach. A handful of statesmen and diplomats are laying the foundations of the new era now. They are hard at work to present the peoples with accomplished facts on the day when hostilities cease. The policies are being laid down now; their execution is being set in motion.

There will be little opportunity or desire, when a war-weary humanity at last hears the cease-fire signal and a sigh of relief goes up from the earth, to reopen vexatious questions, such as Palestine, Burma, France, or Spain. People will want to be rid of it all, of everything connected with the war, and return to the ways of peace. Those who would then seek to open the discussion, or rather ask or suggest that the long-postponed deliberations on postwar reconstruction be officially opened will be advised, if they have not discovered the truth beforehand, that everything has been settled long ago. To that might well be added a cautionary reminder that it is highly inexpedient, as well as dangerous and unkind withal, to plunge society into what would necessarily be acrimonious and rankling debates at a time when men's nerves are still on edge and when humanity, more than anything else, needs peace and calm to recover its breath and composure after the long and sanguinary ordeal through which it has just passed.

At the time when the war had brought us nothing but disillusions and adversities, peace planners were told that discussion of the shape of things to come was premature, if not unpatriotic, in that it would tend to endanger national unity by diverting attention from the one sole objective that mattered: the defeat of the enemy. After the war when the same individuals, organizations, and committees express their concern about the actual state of affairs, pointing to the unfulfilled pledges, the evaporation of the ideals of the Atlantic

Charter, and the menace of new wars, they will be warned with a severity appropriate to the gravity of the then existing situation that discussion and agitation are likely to disturb the precarious equilibrium that has just been established by almost superhuman efforts and that they had therefore better let well enough alone.

When China offers the aid of her armies in reconquering Burma and reopening the Burma Road and is turned down, it means that the fate of China in the postwar world is being settled now. For by that refusal it is made clear that China is not to participate as a free and independent partner in the liberation of Asia from the Japanese usurper and is therefore not to have a say in the postwar reconstruction of the Far East, but that China is destined to be what she was under the *status quo ante bellum,* the re-establishment of which, as Secretary of the Navy Knox said, is the objective of a forthcoming Anglo-American offensive in the Far East: a huge semicolonial area to be exploited by Great Britain and the United States.

When the British High Commissioner in Palestine announces "a vast plan of reconstruction," and this vast plan turns out to be a scheme for reforestation and when at the same time he insists that the White Paper baring Jewish immigration into the Holy Land after the spring of 1944 must stand, then the fate of Palestine not alone, but the fate of millions of Jews in Europe, is not only being settled but it is their doom that is being proclaimed.

"It is through our blind assumption that what happened once will happen again, that we talk today about a peace conference after this war," remarked David ben Gurion, the Palestinian Zionist leader, in an address delivered before the Manufacturers' Association of that country, in June, 1943. "One can say, indeed, that a peace conference is already taking place, while the war is still going on. When Churchill met Roosevelt at Casablanca, that meeting was part of the peace conference. When Eden, together with Halifax, held discussions with Cordell Hull in Washington, that was one of the sessions of the peace conference. When representatives of the Anglo-Saxon world meet representatives of Soviet Russia, this, too, is part of the peace conference. The provisions of the postwar settlement are being planned now, not only in Washington, London, and Moscow, but also in Cairo, Baghdad, Jidda, and Jerusalem.

"And we forget," he added, "that in fact the new order to be introduced in Palestine is absolutely ready. It bears the imprint of British authority, and it is Great Britain which rules in Palestine; and it can be safely assumed that she will rule in Palestine after the war as well. There is a declared British policy, in Palestine. This declaration, unlike others, does not exist only on paper, but is being applied here energetically, consistently and with determination. It is the policy of the White Paper . . . . The peace conference is proceeding in the very midst of war, and the provisions for Palestine's future are also made in the midst of war. The fate of this country (Palestine) is being shaped by the creation of political, military, and economic facts in accordance with the policy laid down in the White Paper: Palestine is now being transformed into an Arab-British state with a Jewish ghetto . . . . "

The moment Germany's desperate attempt to break out of Europe into the traditional sphere of colonialism with its illimitable riches in raw materials had been frustrated at Stalingrad and El Alamein in 1943, and even before Marshal Rommel was at a safe distance from the Near East, the British government took two measures which may well be prototypical of what is to follow if the same old ruthless imperialism and imperial considerations alone are to rule the future of mankind. Without waiting for any peace conference, or even for the final defeat of the Axis, the two eventualities which are always invoked as a pretext to postpone decisions and formulate plans for postwar reconstruction, without debating the matter in Parliament or consulting either the League of Nations, its Permanent Mandates Commission, or any of the nations involved, Britain set out to settle and shape the destiny of great peoples in a way that makes the Atlantic Charter appear, as Bernard Shaw said, as worthless as the paper it is written on.

The two measures in question were the shift in the British government's support from De Gaulle to Giraud and the convocation of a preliminary Pan-Arab Conference in Cairo in May, 1943. If these two measures appear at first glance of a widely different nature and of isolated importance, they are in fact intimately linked, aiming as they do at the selfsame objective, which is the establishment of an exclusive British sphere of interest in the Near and Middle East after the war.

At first, when he broke with the Fascist clique headed by Pétain, Weygand, and Laval which had brought the German army into France to put an end to the Third Republic's democratic regime and announced that he would continue the war by the side of Great Britain, Charles de Gaulle was welcomed with open arms in London. He, the leader of the Committee of Fighting French, which immediately set about to organize resistance to the Axis in the colonies and protectorates of France, was recognized as the representative of the real France, the France that had lost a battle but not the war—a France that would in the end march to victory side by side with the British ally. De Gaulle did not falter or desert. He for one remained loyal to the alliance when so many of his military colleagues defaulted. The English people took him to their heart. The British Government supported him even in the face of American diplomatic opposition, which could not forgive De Gaulle, as Harold Callender of *The New York Times* said, for having spoilt its Vichy policy. This went on for nearly two years, but then gradually the British government, too, grew visibly lukewarm to the French general as it appeared that he had managed to establish contact with the French underground and to have rallied the whole French people, the democratic masses of France, behind him.

For that France, nationalist or revolutionary, represents a latent threat to the imperialist designs to keep intact the vast system of cartels the Germans have built up in occupied France and in North Africa in the last few years. Britain and United States aim to take over these cartels or dominate them in the postwar era.* Charles de Gaulle, compelled by his democratic following, might well destroy them or harness them to a reborn French Empire.

To control postwar Europe and to keep the Atlantic coast of Africa occupied militarily with airdromes and submarine stations in Morocco, Equatorial Africa, and other regions, America and Britain must have control of the French Empire. They dare not take the risk of letting Russia, a Socialist state, share with them in the exercise of power over the Continent once it is liberated from the Nazis. The chief objectives of the United States and of Britain in the war have therefore been to gain control of the French Empire—Morocco, Algiers, Tunis, and Equatorial Africa.

Now Charles de Gaulle and the French underground aim at the institution of a popular, democratic, and independent French regime. Such a regime will of a necessity have to pursue a liberal policy in the French Empire. And that is where De Gaulle clashes directly with Britain and the United States. That is why from being welcomed at first as a hero and a comrade-in-arms in the Allied capitals, he has been besmirched and made to appear a suspicious character, a revolutionary with megalomaniac tendencies, and has even been called a stubborn Fascist. That is why his influence has been more and more curtailed. While the governments of Britain and the United States proclaim that nothing

*See the issues of June July, and August of *The Protestant, A Journal of Affirmation,* 521 Fifth Avenue, New York, on the interlocking German-French-American cartels, a veritable Berlin-Rome-London-Washington Axis.

can be done or should be done in the matter of laying the bases for a postwar order until the Axis is finally defeated, and no bridges should be crossed until we come to them, they are themselves busy, far on the other side of certain bridges, establishing spheres of interests in Africa and Asia and undermining the influence of democratic movements on the Continent.

The break between Charles de Gaulle and the British government came as a result of events in the French mandate of Syria.

For years before the outbreak of World War II the French government had promised independence to Syria. Léon Blum, when Prime Minister of France, actually began to take measures to make that independence a reality. He announced an administrative autonomy for Lebanon and in Syria free elections for setting up a constituent assembly that would bring a Syrian parliament into existence. But as in the case of Republican Spain, which as a sister democratic regime he wanted to succor with war material against a clique of rebellious generals supported by Hitler and Mussolini, the French Prime Minister was persuaded by Downing Street to stifle his generous and righteous impulses.

"Every time," Blum said to me one evening at his home on the Quai de Béthune in Paris, "every time we make the slightest move to help the Spanish Republic, when we want to give them at least an equal chance with the Fascists by supplying them with war material, London brings pressure to bear at once to make us desist, to let events take their course, which of course means to let Fascism triumph. This pressure is accompanied by threats that if France becomes involved in a war with Hitler or Mussolini over Spain, she must not count on British aid . . . .

"And it is the same thing with Syria," he said on another occasion. "Britain does not want us to give Syria her independence. They reason this way in London: suppose Syria becomes independent and in the course of time grows friendly with Germany, enters the Rome-Berlin Axis constellation. That would be a very hard blow to Britain in view of the fact that the pipe lines from the Mosul oil fields pass through Syrian territory, and Syria, at Germany's demand, could close off our oil supply at the termini in Tripoli, in Syria, and Haifa, in Palestine, or in the desert at a moment's notice. Moreover, it is pointed out, a really independent Syria would set a bad example to Iraq and other Arab countries."

"But," said I, "suppose Syria, from being a French mandate territory, becomes an ally of France, as you proposed yourself in the Chamber. France has no intention of withdrawing altogether from that country. She will continue—isn't that your intention?—to exercise a profound cultural influence. France, even in Turkish days, was the protector of the Christians in the Levant and has dotted the countries of Syria and Palestine with schools, eleemosynary institutions and churches. The Vatican, too, looks upon France as the traditional protective power of Latin interests in the Near East."

"Britain does not want a Franco-Syrian alliance either," Blum replied. "That would make France too powerful in the Near East, and she would exert indirect influence on the Mosul oil supply. Britain does not intend to have this come about. She would rather take over Syria herself or add Syria outright to her Arab sphere of interests, perhaps give the country some measure of autonomy in a British-controlled Arab federation . . . ."

In 1941 Britain did take over Syria—that is to say, General de Gaulle did that with his Free French troops in collaboration with the British. And then De Gaulle honorably made his capital mistake: as a democrat and a believer in the sincerity of the Atlantic Charter's promulgators, he immediately gave Syria her independence. This brought on the rupture between the Free French and the British imperialists, who did their utmost to prevent the move. From that moment dated the campaign of intrigue against De Gaulle and the attempts by London to take Syria out of his French Committee's hands and transfer it to the control of Giraud, the imperialists' handy man. With that went a campaign of bribery, the mobilization of the *chevaliers de Saint-Georges,* as the French call English money in the Near East, to have French officers desert the De Gaullist cause for the Giraudist camp. De Gaullist officers in the United States and in Syria are convinced that all this points to the existence of a plan to divide the French Empire between Britain and the U.S.A., under which the whole Near and Middle East is destined to pass into British hands.

Thus Charles de Gaulle, the ally of the first hour, the man who most surely represents the French nation and its democratic aspirations, was sidetracked in favor of Giraud, a military nonentity, whose fame rests upon having managed to have himself taken prisoner twice

. to have escaped twice, the last time under very picious circumstances. But Giraud has this to his credit: he represents the clique of international bankers, landowners, and Franco-German cartelists who have found asylum in Morocco, where they were first protected by the Nazis and are at present under Allied supervision. It is Giraud, and surely not Charles de Gaulle (who showed in Syria that he believes in parliamentary democracy and in letting the people decide their own future), not De Gaulle but Giraud who now, that the French Committee of Liberation is recognized will have under his personal command that new French army, equipped by lend-lease material and composed of Senegalese, which is to take part in the invasion of the Continent. It is Giraud who has been selected to institute in France a regime based on his avowed Nazi doctrines and beliefs that "graduation values are better than political agitation," that "cheerfulness is better than debate," that "there must be a social elite," and that "the army as such has enduring values." If that isn't Fascism, words have lost their meaning . . . .

The second measure, as callously insensible to the national honor and national rights of Israel as the first was boldly contemptuous of the French nation's most sacred aspirations, consisted in convoking, at Cairo, Arab notables from Iraq, Trans-Jordan, and the Hejaz to confer with Egyptian politicians and British diplomatic agents on the future of . . . Palestine. The Jews were off in their tens of thousands fighting and dying in the British Eighth Army in Cyrenaica and Tunisia. In Palestine they were working night and day, in their hundreds of thousands—men, women, and children—for the success of British arms. In Europe the common enemy had slaughtered more than two million Jews whose only hope was to find refuge in their Palestinian homeland, while in the United States the Jewish community, almost to a man, stood with the champions of Britain's cause from the very first hour. Everywhere the most touching and loyal Jewish solidarity with Britain's cause in adversity and prosperity!

And then behind the Jews' back, yet not secretly lest they be spared the humiliation and the cynicism of it all, the British government calls to Cairo the Prime Minister of Iraq, a country that had to be beaten into submission a few months earlier because it had made common cause with the Nazi enemy; representatives of Ibn Sa'ud, the ruler of two million Bedouin, who could not make up his mind where he stood when the British Empire hung on the edge of the abyss; Abdullah, Emir of Trans-Jordan, whose entire kingdom was subtracted from the Jewish people's heritage and who wants still more; King Farouk of Egypt, who was openly pro-Axis, whose Prime Minister was caught *flagrante delicto** betraying British military secrets to Axis headquarters in Tobruk, and whose chief of staff was arrested as he boarded a plane for a rendez-vous with the enemy; and a member of the Higher Arab Committee in Jerusalem, the organization set up by the former Mufti of Jerusalem to fight and terrorize the Jewish community in Palestine and who himself was the Führer's candidate for the post of *Gauleiter* in the

***Free World,* New York, issue of 1942, p. 221, article by Emanuel Neumann on Arab alignments in the Near East.

Holy Land. These gentlemen were called in to say what they thought ought to be done with the Land of Israel and, indirectly, with the Jewish masses in Europe who have seen a thousand Lidices and undergone horrors unimaginable.

Fantastic procedure!

One could readily imagine half a dozen valid reasons and more than adequate justification if the British government had called in the representatives of the skulking Arab tribes who were being whipped up against Britain by the most violent and brazen Axis propaganda when the thunder of the Afrika Korps could still be heard all over the Levant. It would have been understandable, and therefore pardonable, if Britain, in the desperate circumstances attending the battles of Stalingrad, Caucasia, and Libya, had made some offer, some concession, some gesture of appeasement to tribes that had been brought dangerously near the point of explosion. But in May, 1943, when the Germans were in full retreat from their major objectives in Russia and Africa, when hundreds of thousands of Allied troops were being concentrated in Iraq, Syria, and Palestine and when the potential nuisance value of primitively armed tribesmen was reduced to an absolute nil by the presence in their realms of an Allied air force that would have made a brief and macabre interlude of any incipient revolt, in that month and under those circumstances the wooing of these gentlemen in Cairo becomes almost incomprehensible.

Almost, I say, for we did not have to wait long till the Arabs themselves cleared up the mystery. They let the cat out of the bag by their whoops of laughter. A friend of mine, the scion of an Alexandrian family who have been bankers to the Egyptian government since times immemorial, wrote me in August, 1943, that one evening while the preliminary Pan-Arab conference was in progress in Cairo, he was the guest of Egypt's Foreign Minister. With him among the invited were several foreign delegates, one of them His Majesty Abdullah ibn Hussein, Emir of Trans-Jordan. When the coffee was being served, the conversation turned to the affairs of the conference, and my friend said that its convocation by the British agency had come to him as a total surprise.

"I thought," he said, "that a long time would have elapsed before the British government would have found the time and the inclination to begin a discussion of Near Eastern affairs." To this the delegate of Palestine's Higher Committee replied: "Yes, we in Jerusalem, too, were surprised to be called to Egypt. Some of us figured that we might well be sent to the Seychelles Islands.* Personally I rather expected to be imprisoned upon the defeat of the Germans. But then, *ashkor Allah,* Allah is merciful and does not deal with us according to our transgressions. Here I am, the honored guest of His Britannic Majesty!"

"At this remark," relates my friend, "Abdullah chuckled till the tears rolled down his cheeks, and the whole company burst into uncontrollable fits of laughter in which I joined heartily!"

And well they might laugh!

---

*Some Arab terrorists were banned to these islands during the disturbances of 1936-1939.

While thirty odd thousand Palestinian Jews were fighting in the British Eighth Army and another one hundred thousand young Jewish men and women had registered for military service and the whole of Jewish Palestine worked night and day for Britain's victory, the British administration in Palestine concentrated its attention on immigration—that is to say, it fought tooth and nail against the admission into Palestine of individuals who by almost superhuman efforts had managed to escape the place where, according to the Polish government-in-exile, hundreds of thousands of Jews have been slain and where the slaughter continues. In fact, even before the war, ever since the European situation started deteriorating, Jews began escaping to Palestine by sea and land. They chartered ships that put them ashore on deserted beaches at night. They came in small boats and big steamers. In one case, that of the Turkish steamer *Sakarya,* which arrived off the shores of Palestine in February, 1940, as many as twenty-four hundred Jews succeeded in landing. They were all refugees, without passport or visas and were therefore considered and treated as illegal immigrants.

Aroused by "this systematic infringement of the law," the government had the beaches patrolled. Shooting at the boats, arresting and interning those who were caught trying to come ashore, confiscating their ships, giving long terms of hard labor to the captains and crews (who were mostly Greeks), and jailing the Jewish organizers of the "undersea lane" were among the lesser tragedies.

There were also tragedies of a much more serious

character brought on by the Palestine authorities. For instance, in the fall of 1940, several small ships bearing refugees were apprehended off the coast of Palestine. The nineteen hundred Jews on board were transferred to the large liner *Patria* in the port of Haifa. Despite extensive demonstrations of protest and strikes throughout the country, the government decided to exile the refugees to the island of Mauritius, near Madagascar. In a letter smuggled off the ship one of the refugees wrote to a friend in Haifa: "We have decided to remain in Palestine, dead or alive." And so they did. For on the morning of the day when the *Patria* was due to sail, an explosion occurred. The ship capsized and sank to the bottom in ten minutes: 150 of the nineteen hundred passengers perished. The catastrophe might have been smaller, for it so happened that the explosion occurred while all the refugees were lined up on the decks to be counted, but the British captain of the ship, hearing the detonation and thinking that there was an air raid, ordered everyone into the holds . . . . The survivors were granted permission to remain in the country.

About the same time two thousand more Jews were caught trying to enter Palestine without visas and were placed in a concentration camp at Atlith. One night the two thousand were ordered to get out of bed and be ready for transportation within an hour. They refused to obey; instead, they all undressed and lay completely naked in their bunks—men, women, and children. At two in the morning, tired of negotiating, the police, aided by the army, routed them out of their huts and loaded the men, women and children—naked as they were—into trucks, giving each one a blanket, and by sunrise they were aboard Dutch steamers on the way to Mauritius. Those who resisted were mercilessly beaten. The floors of the huts in Atlith, reports Sir Victor Gollancz,* were red with blood the next morning. In Mauritius, no authorized Jewish representative was permitted to visit the concentration camp and very little is known about the condition of the exiles in the Pacific. Mauritius is known for its beautiful birds and famous leprosarium.

A third tragedy occurred a year later, at the moment when Raziel** and his Jewish volunteers were busy saving the British oil refineries of Mosul from destruction by the Iraq rebels. In the atmosphere of terror in Rumania, where the bands of Anton Ionescu, Hitler's friend and ally, were slaughtering Jews indiscriminately in Bucharest, Jassy, Braila, and Galatz and the number of victims was reported to exceed 100,000, the news that the *Struma,* a 180-ton cattle boat, was to sail to Palestine found swift currency. But only a few could find the money demanded for a passage to Palestine. The Rumanian Ministry of Labor put obstacles in the way of the sailing, but eventually the *Struma* left Constanza on Friday, December 12, 1941. Custom officials stripped the passengers bare of most of their clothing and bedding and of all foodstuffs which the refugees had brought with them. None was allowed to take more than forty pounds of luggage. There were no medical supplies on board. A Rumanian boat piloted

**Let My People Go,* by Victor Gollancz, London, 1943.

**See *That Day Alone,* Garden City Press, New York, pp. 337 *et seq. "The Road to Baghdad."*

the *Struma* through the mine-strewn harbor.

Very soon it became clear that the engine was old; it had, in fact, been taken from a boat sunk in the Danube and several years under water. The ship's mechanics were incompetent. The boat made frequent halts. Neither the wireless nor the projector was in order, and both failed to work. There were no lifeboats and only a handful of life belts.

After repairs the vessel groped along the coast until it reached another Rumanian port, where the captain of another steamer tried to repair the engine—against payment of any possessions left to the passengers. The next day, in the Bosphorus the engine failed again, and the boat was unable to make any further progress. The boat carried 769 passengers, including seventy children under thirteen years of age and 250 women.

On board there was not even enough room to move freely during the day, let alone sleep at night. The passengers were given quarters in cages built along the ship's sides, each of the cages, which were five feet wide and eighty inches high, being occupied by five persons. There was no room for any water supply. There were no sanitary arrangements except for one water closet on the deck. There was neither electricity nor heating. The thirty doctors—passengers—had not even one bottle of disinfectant among them.

The space below decks was foul with coal fumes, while the deck provided just enough space for some of the passengers to stand motionless, which they did by turn. Only the intense cold prevented an epidemic, threatened by the insects swarming on board. Dysentery ravaged the passengers. Several lost their reason.

The trip from Constanza took four days instead of fourteen hours. Only the hope that somewhere on the Turkish coast they would be mercifully welcomed sustained the refugees.

Arrived at Istanbul, neither the crew nor the passengers were allowed to come in contact with anyone outside the boat. The representative of the agents notified the passengers that he had no money either to repair the engines or to supply food for the passengers. For eight days the boat was anchored at Istanbul, after which the passengers managed to obtain a daily ration of food. It was impossible to get any fuel in Turkey, and the passengers could consequently not cook any food.

The refugees soon learned that they would not be permitted to disembark so as to make their way overland to Palestine and that the engine must therefore be repaired. When this was begun it became apparent, however, that it would be impossible to finish the job before the end of January. The Turkish authorities permitted only restricted purchases of foodstuffs. Without air, without light, without the means to wash, or eat, or rest, passengers were constantly ill, and the doctors worked tirelessly.

The rest may be told in the sober words of a statement issued by the executive of the Jewish Agency for Palestine:

> The Jewish agency has learned with grief and horror of the sinking in the Black Sea of the SS *Struma,* with 750 Jewish men, women, and children aboard, refugees from Rumania. The boat had been lying at Istanbul since December 15,

and the conditions as regards food and sanitation were described as "desperate." Every effort was made by the Jewish agency in Jerusalem to persuade the Palestine administration to admit these victims of persecution. The agency proposed that they should be put to the account of the immigration schedule which permits three thousand Jews to enter Palestine during the current six months. The matter was also taken up with the Colonial Office in London. The only concession obtained was in respect of children under sixteen who were to be allowed to enter Palestine.

Two sets of arguments were put forward by the British authorities against granting the request of the agency to make adult refugees on the *Struma* also eligible for certificates under the current schedule: first, that these people had been under the Nazis, and that they might therefore include some enemy agents; secondly, a shortage of supplies in Palestine.

As regards the first point, the agency has repeatedly urged that such refugees should be placed in internment camps and not released until and unless characters were established to the full satisfaction of the Palestine administration.

As regards the second objection, more than two thirds of the *Struma* refugees were people fit and willing to work and to fight. To use the argument of "short supplies" against admitting some two hundred elderly people flying from torture and death reflects on the intelligence, as well as on the heart, of those who advise it. Moreover,

> as these people would come under the schedule already granted, the problem of supply did not arise . . . .

After all attempts to negotiate had been exhausted, the ship turned back towards Rumania and was blown up by the act, it is believed, of the passengers who preferred death to return. It may be mentioned that the concession permitting the children under sixteen to enter Palestine arrived too late—after the ship had sunk . . . .

What then is at the bottom of that strange policy of delay, obstructionism, and yes, even cruelty, which the Colonial Office has pursued with regard to the Jewish national enterprise almost from the moment that Great Britain was entrusted by the League of Nations with the administration of Palestine?

Is the presumption perhaps justified that Britain for some reason or other regretted having liberated the Near East from the tyranny of the "unspeakable Turk" and of having undertaken, at a moment when idealism ran high, the institution of a new, more humane, and civilized order in the Arabian Peninsula?

Have the Jews done anything to forfeit the confidence of Great Britain or the confidence of the other states which, by ratification of the mandate, became cosponsors with Britain of the undertaking to provide a homeless people with a home? Was there anything in the feeble beginnings of the Jewish national home in Palestine, something in the mere outline of the political and economic structure slowly arising in the Holy Land, that led to a reconsideration of Britain's initial program? Had the local administrators perchance recognized in the embryonic Jewish national undertaking certain elements of danger, which would, in their opinion, if allowed to develop unchecked, prove to be a potential hindrance to British imperial policy in Iraq and Egypt in later years or likely to have a detrimental effect on the British Empire's interests? Something like that may explain why the Colonial Office, as soon as it got its hands on Palestine, began systematically limiting, circumscribing, whittling down, and minimizing the scope of the Jewish national home.

When the Colonial Office took over from the War Office and the Foreign Office, it disregarded, first hesitatingly, then more and more boldly the home government's previous commitments. It instituted a definite policy of obstruction. That policy originated among British Near Eastern political and military circles. Within a few years after the liberation of Palestine from the Turkish yoke and the assumption of a mandatory power's duties by the British Government, that benevolent spirit which was evident in the Balfour Declaration and which manifested itself supremely at the time of the San Remo Conference, where Britain formally undertook the Mandate, suddenly evaporated. What brought this about?

It is inconceivable that enlightened Englishmen should suddenly have come to the view that "the glory of the divine justice delights in the sufferings of the Jews" and that the Jewish people for that reason must forever, as if by a providential fiat, continue as wanderers in the earth in order that God be kept in good humor or that the historical authenticity of certain curious details of the Gospel story be validated by their enduring misery. Nor would anyone charge that the British government anticipated Adolf Hitler in adopting an anti-Semitic policy overnight. What condition then, or event, or eventuality looming on the future's horizon caused Britain to shy away and progressively to swerve ever farther from what the world in general, and the British and American peoples in particular, took to be her generous earlier intention of turning Palestine into a national home for the Jewish people and thus of making an attempt to end the scandal of history of a great and ancient people being compelled to haunt the corridors of history as ghosts and beggars and to roam about as waifs of every storm that blows? Yet to that end, and to no other, did the Christian world accede to Britain's demand that she be given charge over Palestine and there institute such conditions as would lead speedily to the long overdue national emancipation of the Jewish people.

Men like Arthur Balfour, Smuts, Masaryk, Lloyd George, Winston Churchill (before he became Prime Minister), Croce, Woodrow Wilson, Colijn, Ragaz, Nansen, Guignebert, Herriot, and Wyndham Deeds, the intellectual, religious, and political elite of every nation, each in his own way and in virtue of his own, often widely divergent philosophy of life, justified the redemption of the Holy Land by the Jewish people as the supreme remedy, the surest instrument for a solution of the Jewish problem. By contact once more with their own soil whose nourishing and creative emanation had been denied to them for long centuries the

Jews were to cure themselves of the maladies and abnormalities the Dispersion had inflicted on them. By setting up a national workshop or laboratory so to speak, where Jews could once again, as in the past, in their own name and in the spirit of their own national ethos, work out their own national Hebraic conceptions of what man's relationship with his fellow should be, they were to have their shattered nationhood restored. Untrammeled by the fear of persecution which their ineluctible nonconformity visited upon them again and again in the Dispersion, unfettered by the paralysis of the ghetto mentality which produced an inferiority complex, self-debasement, and self-hate and which stifled the creative spirit, free from the molds and straightjackets of this or that alien civilization, and in short, by taking up a common, collective national task, it was hoped—and that hope has gone far towards fulfillment—that the Jews would recover from the sickness of their national homelessness, which is the malady that destroys them spiritually and physically in the Dispersion, with its demands of assimilation and abandonment of national values and character. Thus, it was hoped would come to pass what is said in the Book of Job: that the flower resuscitates by the smell of water and the scattered Jewries in the world receive the revivifying spiritual currents from a regenerated homeland to refresh their drooping and withering life....

Can nobler task be imagined than that which devolved upon England: to serve as the instrument for the healing of a people's soul and to provide a haven of refuge for the victims of racial persecution?

The Englishmen of Balfour's day knew that whoever takes up the cause of the Jewish people touches upon eternity and that it is not might of arms but righteousness alone that exalts a nation. Why then did the British government permit every trifling argument that opponents of Jewish national emancipation and its chief instrument, the establishment of a Jewish National Home in the Jewish land, could invent and bring forward to grow into mighty and formidable objections? And why do British statesmen till this day seize with an almost perfervid eagerness upon every show of hostility to the Jewish national idea, from whatever direction it may come, be it from misled Arabs, from vulgar Jew-baiters, or from the upper strata of the Jewish *bourgeoisie,* as welcome ammunition in the Colonial Office's campaign of blasting Jewish national hopes? Why is there now this attempt to "freeze" Jewish endeavor in the Holy Land and curtail the growth and development of the national home to the shape and status it has reached at present? Why was the first generous and just impulse so soon abandoned?

Can it be that the permanent officials of the Colonial Office have elaborated another plan for the Near East, that they have another political arrangement in view for that part of the world where the Second World War has eliminated all of Britain's actual and potential imperial rivals? Is the process of transforming Palestine into a national home for the Jewish people to be arrested under this new plan or to be abandoned entirely in favor of an Arab federation, under Britain's aegis, in which Palestine is to be included? Will this new plan make of the existing Jewish community an infinitesimally small minority, not even an enclave in the Arab world, but a ghetto where Jews will be outnumbered twenty-five to one and will thus automatically lose their voice, see their initiative paralyzed, and be compelled to dance to the tune of Arab overlords and to assimilate themselves to a primitive civilization?

The answer to this question is that it looks very much like it. It is true that the plan is not clear in all its details. Only the general outline has become discernible. The British people are by no means aware of all its implications. In England, as in America, the notion, which was inculcated by the propaganda of the Colonial Office, still prevails that all the trouble and friction in the Near East derives fundamentally from a congenital Arab-Jewish incompatibility and that any solution therefore is better than the present uncertainty and the actual situation's explosive nature.

It is true that in so far as the Colonial Office is concerned the full import of this plan cannot be disclosed without at the same time disavowing, to which the British people would never consent, the terms of the Atlantic Charter and thus depriving millions of human beings, not only the Jews, of the ideals and hopes that sustain them in and through the terrible ordeal of war.

When Henry Wallace, that valiant member of the *Militia Christi,* refers to the end of imperialism and of the need for Occidentals to begin thinking in terms of the equality of races, there is no answer in imperialist circles, neither in Britain nor in America, no sign of recognition or understanding, neither a word of assent or dissent. Plenty of generalities, of course, plenty of platitudinarian repetition of Atlantic Charter phraseology, but no going to the root of the matter, no frank and open and fearless deliberation, at least never in public, on what the shape, the form, and the spirit of that better world of the future is to be.

What kind of freedom are the United Nations fighting for? Economic freedom? Racial freedom? National liberation and independence? Economic and racial equality? The war in the Far East, where almost the entire population of a billion people live as colonial subjects of the Great Powers, was fought under the ideological and military concept of conquerors. No offer of freedom was made to the colonial subjects of the empires engaged. No arms were given to the people to increase the possibility of their support in resisting the Japanese aggressors. Imperialist prerogatives ruled the conduct of the United Nations in Asia. The results were: Burma lost, Malaya lost, Singapore lost, the Dutch East Indies lost, India in turmoil, Australia threatened. The question may well be asked: would this have happened if the Allies were fighting a war of genuine freedom for the colonial peoples? Would the peoples conquered by the Japanese show that contemptuous indifference to the Allied cause they continue to exhibit, and would Japan make the enormous headway she is making in converting the subdued races to her concept of Asia for the Asiatics and the Eastern Asia coprosperity Sphere?

Lord Lyttleton, British Minister of Production, assured the world that "we have passed from an age of scarcity to an age of plenty." Viscount Cranborne,

Secretary of State for the Colonies, stated in the House of Lords on June 2, 1942, that the Atlantic Charter lays down the fundamentals on which a peace settlement is to be based. But he went no further. "Anything said today," he went on, "would not merely be useless, but more likely to do more harm than good!" In other words: an explanation or a discussion of a fundamental question, not some technical detail or other, but of a fundamental question over which the world is in arms, for which a score of millions have already laid down their lives and in the struggle over which one house in every five in Britain lies in ashes, would do more harm than good! Harm to whom? Is Hitler not to know what the Atlantic Charter means? Or are the peoples who make the sacrifices, whose wealth is being destroyed and whose lives are forfeited, not to know what is in store for them? Is their fate a nice little surprise that is being kept in some magic Pandora's box till the end of the war by their lordships, the merchant princes, bankers, and monopolists in England?*

A great British newspaper quoted Sumner Welles as saying that after the war the nations of the world must have "free access" to raw materials and that "the United Nations will provide the mechanism whereby what the world produces may be distributed among the peoples of the world." The British newspaper enthusiastically agreed with the American statesman. Did Britain then not have "free access" to the world's raw materials before the war? And what prevented Britain from making Indians, Malayans, and Burmese, France from making

*See the National Question in Europe, New International, Feb. 1943.

Indo-Chinese and Algerians, and Holland from making Indonesians "free from want" by distributing among them what the so-called mother countries produce? What is the marvelous new device the United Nations are going to provide? Until now, the distribution of what the world produces has been carried out by trusts, monopolies, and banks. It has led the world into economic crises always more profound, always more persistent, to tensions ever more acute, and finally to a war from which it becomes ever more difficult to recuperate. What is the new mechanism that the United Nations are to provide, the new kind of distribution? Why aren't we to know? Is that secret too, to be kept from Hitler?

The thing that stands out in speeches and statements of the Allied leaders who take up the question of the postwar era is that they are filled with generalities. And the reason is clear: to match their phraseology with concrete illustrations of their general theories, to present a true program of economic, social, and political democracy, would mean to show up the failure of the old capitalist mode of production as a world social order.

The only plausible argument for the colonization of those areas in the globe inhabited by the so-called backward peoples—the Indians, Indonesians, Malayans, Chinese, Africans, and Arabs—who constitute the greatest proportion of the human race was to lift them out of their historical stagnation by bringing them the benefits of modern capitalist civilization. The fulfillment of this task was the famous "white man's burden." But it was at the same time to assure the establishment of markets for capitalist production. In other words, civilization's task was to work both ways: it was to bring prosperity at home and at the same time improve the lot of the colonial peoples by lifting them to a higher cultural *niveau* and to a higher standard of material well-being. In theory a virtuous enough arrangement, but in practice it turned into drawing off the tropical products and industrial raw materials in which the colonial areas abound without developing those countries industrially so that they retain a predominantly agricultural character even in an industrial age. Imperialism showed little concern about industrializing and advancing the productive capacities of its colonies by introducing modern, heavy industry, but appeared mainly concerned with exploiting those areas and reaping the profits thereof.*

What this has meant in the course of time to the peoples living in distant colonial territories may perhaps best be seen in events which have occurred but recently much nearer home and in full view of the entire civilized world—namely, in Germany's belated appearance on the scene of history as a colonial power. Finding the choicest lands of the earth—that is, countries abounding in raw materials—occupied by others, denied access to these resources, unable to seize them by force of arms, the Reich proceeded in the most ruthless and reactionary manner to create herself a colonial empire in the only area lying within immediate reach of her armed forces: Europe.

On two occasions within the short space of a quarter

*See Fourth International, Feb. 1943.

century, the German Reich, desperately in need for markets, struck out but failed each time to break out of the confines of Europe into the classic lands of imperialist colonization. Held back by force, the Reich was forced to content itself with establishing its colonies in Europe and extending its sway, not over historically backward peoples, but over nations living in the heart of Western civilization. These she reduced to absolute slavery. Being denied access to the raw materials of the world and yet intent on maintaining an untempered capitalist mode of production, Germany set out on that campaign of conquest, beginning with those smaller states which made up the ring of steel her rivals had forged around her after her first attempt to break out.

And then a most curious development took place: the government of the Reich announced and instituted an economic policy for the New Order in Europe under which the conquered countries were to be deprived of their industries and were to function solely thereafter as agricultural hinterlands for Germany and German industry. The Reich, the so-called German *Herzland,* was to be the great industrial center of the Continent. The *Herzland* alone was to have an air force so that revolts in the "colonies" would for ever be impossible. The rest of Europe was to serve as a market for German products; its peoples were to be Germany's slaves. What the white man's burden had been once in Asia and Africa was to become, but now accompanied by nameless cruelty and a total disregard for human values, the prerogative of the self-styled superior Nordic race. The conquered peoples in Europe were placed in a position of legal inferiority to members of the triumphant *Herrenvolk,* and thus the institution of extraterritoriality, accentuated to its ultimate consistency, was brought to the heart of Western civilization at a moment when it was progressively being abolished in Asiatic and African lands of colonial and semicolonial status. The European peoples were reduced to the rank of "natives," a circumstance that logically involves the assumption by their conquerors of the rank and role of lords of creation.

In carrying out this new policy, the Germans, being in a hurry and under the immediate and constant pressure of war, behaved with an utter disregard for human life, historical traditions, and cultural institutions in their attempt to remake Europe on the model of what Asia had been at its worst. They closed the national universities of the conquered peoples. In order to eradicate national consciousness in the subjugated nations, they killed off the intelligentsia and cultural leaders and warred on the churches. German was made the language of instruction in the schools where schools were permitted to exist. They did not hesitate to transplant whole populations and to exterminate others who could never be expected to assimilate to the new conditions of Nazi overlordship. By separating men and women in Poland, by sterilizing Czech peasants, and by keeping two million Frenchmen in prison camps, they depressed the birth rate of nations considered ultimately unabsorbable in the Nazi order. Everywhere they destroyed working-class organizations with a ruthlessness that indicated their special concern with everything that had to do with industry. But they also destroyed the native capitalisms of Czechoslovakia,

France, Belgium, Holland, Poland, Greece, Norway, and Jugoslavia in the interests of German capitalism.

This phase of Germany's war policy, the barbarous repression of the conquered peoples of Europe, which, more than anything else, aroused the indignation of the civilized world, is remarked here because in principle and in practice it has for a long time—and without the pressure of war—been the identical policy imposed upon the "native" peoples in the colonial countries by the imperialist countries. The Germans destroy native industries in their European colony; Britain, France, and Holland do not permit a modern native industry to come into existence in their colonies and protectorates. And they do it for the same reason: to enrich the governing class of bankers, monopolists, and merchant princes.

The inhuman treatment meted out to the peoples of India by British imperialism is the most striking case in point and has a direct bearing upon and connection with the future of Palestine and of the regions peopled by Arabs. It is Fascism, unmitigated and untempered; a long story of brutalities, indignities, and enforced poverty heaped upon races and nations that were civilized and had produced great art, philosophies, and religions when the ancestors of the English and the Dutch still dwelt in caves and drank mead from the skulls of their enemies slain in the Teutoburger Forest.

To all demands, pleas, and efforts on the part of the Indian peoples to be allowed a say in their own affairs and to develop their immensely rich country and thus wipe out the poverty (which one Englishman described as so harrowing that he would not be surprised if the people turned to cannibalism), to all appeals for a wider measure of freedom, the British government has turned a deaf ear and covered up her immoral and unjust policy by creating false issues and injecting false arguments.

The Indian people were told they had to submit, submit forever to humiliating and bitter bondage. This order Gandhi and the people of India countered with a decision to resist, not by brute force but by what is called *satyagraha,* "soul force." They accepted suffering instead of inflicting suffering in retaliation. That choice of the Indian people was in the spirit of the Christ whose temples stand in England. It was a choice, a decision, which no other people ever made; no people can ever make a nobler one.

How did Britain meet this decision of India to undertake the winning of her freedom and free association with Britain by purely moral force? Has she met it on moral grounds, by considering moral principles, by acknowledging India's moral right to freedom? "No," says Dr. Jabez Sunderland,* an Englishman born, who was for years the representative of the British Unitarian Society in India. "She has ignored and trampled under foot the question of moral right or wrong, and has met India's actions by pure brute

**India in Bondage,* Jabez T. Sunderland, M.A.L.D., Lewis Copeland Comp., New York, 1932.

force. In 1937, under Lord Irwin [the present Lord Halifax, British ambassador to Washington] she met it by issuing and enforcing a series of repressive ordinances so severe that they shocked the lovers of freedom and justice in all countries. And later, under Lords Willingdon and Linlithgow, she has met it by ordinances still more shocking and terrible.

Great Britain claims that she is not waging war on India as Germany is on Holland, Poland, and the other countries in her European *Lebensraum*. What justification could she indeed have to wage war against a people who have no guns, no swords, no airplanes, no arms of any kind, whose only desire is to gain freedom, to see their land developed and the thrall of misery broken and who are endeavoring to win it without bloodshed or hatred, and who would rather lose their own lives than take the lives of Englishmen?

But if the British are not waging war, Dr. Sunderland asks, what justification have they for forcing on the people ordinances and inflicting on her cruelties, brutalities, and atrocities which are quite equal to those of war . . . and these in time of peace?

The ordinances of Lord Irwin, the present Lord Halifax, were so drastic; the atrocities they caused so shocking, that they attracted the attention of the world and were severely criticized. State legislatures in America and the Congress of the United States condemned them* as "acts of violence, infamy and inhumanity."

This was in 1930, three years before Hitler came to power, when newspapers still asked their correspon-

*By resolution in the Senate: July 17, 1930

dents in Europe why the man with the Charlie Chaplin mustache made so much fuss.

Lord Halifax's ordinances were in force for one year. They were followed by a "truce" during which the Mahatma went to Europe to attend the Round Table Conference. In his absence Lord Willingdon, Halifax's successor, broke the truce and issued a series of ordinances even more sweeping and ruthless, if possible, than those of his predecessor.

Hitler was still not in power. German Fascism had not yet triumphed over the German people. When it did, Hitler merely needed to copy the British ordinances in India in order to have a perfect and complete Nazi legislation and a set of rules for a Nazi way of life.

The Indian ordinances ruled among other things:

1. Anyone suspected may be arrested any time, without warrant, and held in jail without trial for an indefinite duration of time.

2. Special tribunals of three persons were to be set up to try political offenders.

3. Verdicts of two persons of the three were to prevail.

4. The tribunals were to meet and carry out their proceedings *in camera.*

5. The accused were not to be allowed witnesses or counsel or any form of defense.

6. Accused persons were allowed to be tried in their absence.

7. The tribunals could give sentences of death or of deportation for life.

8. No appeal was allowed from the decisions of the tribunals.

9. Proceedings of the tribunals were to be secret, no records to be kept, no written notes or other writing allowed, everything was to be oral (to prevent information from getting to the public.)

10. The government was empowered to confiscate any property, movable or immovable, land, houses, furniture, vehicles, retaining the same permanently or as long as it sees fit.

11. Any person or persons could be ordered by a government official to reside or not to reside, to remain or not to remain, in any designated place or area; and also to conduct himself or themselves in such a manner, or to refrain from such acts or activities, as may be designated by the said official.

12. Various classes of persons—teachers among others—could be conscripted for government service and compelled to spend as much time and render as much aid as required, in maintaining law and order, protecting government property.

13. Private and public conveyances and all means of transportation by railroad, trolley, bus, automobile, or public or private carriages of any kind could be denied to any person or persons.

14. Collective fines were to be imposed on villages or other communities in case of certain offenses committed by individuals.

15. Whole villages could be punished on suspicion that persons in the village had harbored suspects.

16. Boycotting, however peaceable, was declared a crime.

17. An attempt peacefully to persuade any person not to buy liquor was made a crime punishable with two years' imprisonment with hard labor.

18. The National Congress, local congresses, committees, or organizations connected in any way with the National Congress were declared illegal and suppressed. Lord Willingdon suppressed 554 societies in Bombay alone, including social clubs, youth leagues, student societies, clubs, women's organizations, knitting circles, and charity societies.

19. Any money, securities, or credits believed by any local government to belong to or to be intended for the use of any unlawful organization, such as the National Congress, may be seized and forfeited to the Government. (The moneys, securities, etc., may not have anything to do with any unlawful association, but no appeal is allowed and no defense is allowed to be undertaken by the injured parties).

20. A severe censorship was placed on mails, telegrams, cables, including letters and reports sent to English and American newspapers. The result is that India, a subcontinent, inhabited by 500,000,000 people, directly in the path of Japanese aggression, was, as far as the the United States press is concerned, *spurlos versenkt.*

21. The local governments were empowered to require grocers, or other tradesmen to refuse to sell anything, even medicine and food, to persons suspected of having any connection with the National Congress or any organization, like a knitting circle, disapproved by the government . . . .

These ordinances have been and are being carried out in peace and war in a spirit of iron determination to crush the Indian National Congress, to crush the Gandhi nonco-operation movement, to crush the whole movement of the Indian people for real dominion status, for self-determination, home rule, for real freedom in any form to be carried into effect within any near or definite or even discernible future. "These ordinances have deprived everybody in India," said Sir Abdur Rahim, a distinguished Moslem judge, "of every vestige of right: right to personal safety, right to liberty, right to association, every kind of right that every human being may possess."

Ruthless ordinances, issued for a ruthless purpose and enforced by half a million ruthless policemen and a strong army equipped with all the modern appliances for war, against an unarmed and peaceful and law-abiding people, have produced terrible atrocities in their enforcement and have turned India into a house of tears and mourning.

For what? To educate the Indians to self-government, as the Conservatives cynically explain? The chief issue involved in this actual or latent state of warfare, this Fascistic state of affairs in India and in the world's colonial areas, is that the colonial powers object and resist to the point of violence the development of the resources of their colonies by the native populations. Men like Churchill, Amery, Atlee, and Halifax in Britain, Herriot, Daladier, Paul Boncour, and Blum in France, and De Geer, Gerbrandy, and Alberdsma in Holland are not anti-Indian, anti-Tonkinese, or anti-Indonesian by nature or inclination or by virtue of some doctrine of Western superiority. Most of them are Christians and make profession of belief in the brotherhood of man. They would indignantly reject any allegation of fostering racial discrimination. They are, when you hear them, quite solicitous about the welfare of the natives in those distant lands, and they support missions, educational programs, and campaigns to improve the cultural, hygienic, and economic status of those peoples. But they also place considerations of imperialist and class policy before moral, ethical, and religious considerations, or as Lord Halifax phrased it once: "At times moral considerations must give way to imperial policy."

They will not permit the industrialization of the colonies because they believe that it would harm and ultimately cause the ruin of the system of production in their respective homelands.

Jesus said: "where a man's treasure is there his heart shall be also!"

All the strife and friction between colonial peoples and colonial powers revolve around the one single issue of industrialization.

For that reason, to keep India a predominantly agricultural continent, to hold on to India, which is the source of the wealth of the British ruling class, Britain has sought control and has acquired control of the entire road to India from the homeland by sea and by land. For that reason she has established herself and maintains herself in such (to her) relatively unimportant countries as Egypt, Iraq, and Palestine. These countries are mere accessories to India, mere stepping-stones on the road to India, supports of Empire.

Not because of Arab opposition (the Arab princes are but the tools of British imperialism) will the Jews be denied their free homeland, but because of imperialism, and as long as it survives in its present form. And the peoples of Asia will never have real political power so long as it is the declared policy of the United Nations to reestablish the imperialist *status quo ante bellum*. Because these peoples in India, in China, Malaya, Indo-China, and in Indonesia know this, they will not co-operate in resisting Japan but will leave the reconquest of their countries to be paid for by innocent American and English blood . . . .

And I (for this must also be said), in writing of these things, am not injecting a note of discord into Anglo-American unity, nor am I obstructing the war effort. I am doing the very opposite. In that coming struggle for defeat of the Japanese usurper, I want to see American and English lives saved by gaining as allies the teeming millions of India, China, and Indonesia, who live in the area of prospective conflict, who do not have to be transported across immense Oceans, and who are willing and eager to stand by our sides if we will but let them know that their co-operation is not merely desired to resist the aggression powers but that it is the great turning point in their own struggle for freedom, their dawn of liberty.

In Palestine, too, the matter of modern industrialization is at the root of all the unrest, dissatisfaction, bloody upheavals, and three-cornered animosity of the last twenty-five years, which will soon again lead to an explosion. The objections raised against the Jewish national enterprise change in depth of ignorance or infamy according to the locale of utterance or origin: the Y.M.C.A. in Jerusalem, the House of Lords in London, the editorial office of *The Christian Century* in Chicago, the Iraq Legation in Washington, Ibn Sa'ud's harem in Taif, or the American Jewish Committee. These objections are that the Zionists rob the Arab peasant of his land, that they want to rebuild Solomon's temple on its old site now occupied by the mosque of Omar and institute a rigid theocracy, that they endanger and desecrate by their presence the Moslem and Christian holy places, that they are agents of British imperialism in disguise and nourish secret imperialist ambitions of their own, or that they are rabid nationalists who go counter to the universal religious mission of Israel. All these charges and allegations over which so much blood and tears have been shed are not only untrue, they are beside the point. They have nothing to do with the real issue at stake except that—as is the design—they obscure it and divert attention from it.

You can say what you like about Anglo-Jewish-Arab tension in the Near East, about its sources, origins, and causes, and even go to the extent, as Lord Wedgwood did in the House of Lords, of declaring that the British Palestine administration is in the hands of anti-Semites pure and simple: it all leaves the Colonial Office indifferent. That institution, the central organ of British imperialism, pursues its way unruffled as long as the real object and cause of its opposition to the Jewish national home are not revealed.

The White Paper of 1939, which bars further Jewish

immigration into Palestine after March, 1944, and which is the triumphant culmination of the sly, surreptitious policy pursued by the antediluvian die-hards of the Colonial Office since 1920, must be made to stand by hook or by crook. In 1939 Winston Churchill called it a "plain breach of promise, a repudiation of the mandate. I regret very much," he said, when the document was discussed in Parliament, "that the pledge of the Balfour Declaration, endorsed as it has been by successive Governments, and the condition under which we obtained the mandate, have both been violated by the government's proposals . . . .

"Is our condition so perilous and our state so poor," he asked, "that we must, in our weakness, make this sacrifice of our declared purpose? Although I have been very anxious that we should strengthen our armaments, and spread our alliances and to increase the force of our alliances and to increase the force of our position, I must say that I have not taken such a low view of the strength of the British Empire or of the very many powerful countries who desire to walk in association with us; but if the government, with superior knowledge of the deficiencies in our armaments which have arisen during their stewardship, really felt that we are too weak to carry out our obligations and wish to file a petition in moral and physical bankruptcy, that is an argument which, however ignominious, should certainly weigh with the House in these dangerous times. But is it true? I do not believe it is true. I cannot believe that the task to which we set our hands twenty years ago in Palestine is beyond our strength, or that faithful perseverance will not, in the end, bring that task through to a success. I am sure of this, that to cast the plan aside and show yourselves infirm of will and unable to pursue a long, clear and considered purpose, bending and twisting under the crush and pressure of events, I am sure that this is going to do us a most serious and grave injury at a time like this . . . .

"What will the world think about it? What will our friends say? What will be the opinion of the United States of America? Shall we not lose more—and this is a question to be considered maturely—in the growing support and sympathy of the United States than we shall gain in local administrative convenience, if gain at all indeed we do?

"What will our potential enemies think? What will those who have been stirring up these Arab agitators think? Will they not be encouraged by our confession of recoil? Will they not be tempted to say, 'They're on the run again. This is another Munich,' and be the more stimulated in their aggression by these very unpleasant reflections which they may make? . . .

"Shall we not undo by this very act of abjection some of the good which we have gained by our guarantees to Poland and to Rumania, by our admirable Turkish alliance and by what we hope and expect will be our Russian alliance? You must consider these matters. May not this be a contributory factor—and every factor is a contributory factor—by which our potential enemies may be emboldened to take some irrevocable action and then find out, only after it is too late, that it is not this government, with their tired ministers and flagging purpose, that they have to face, but the might of Britain and all that Britain means?"

Sir Archibald Sinclair, Secretary of State for Air, denounced the White Paper as a grave departure from the terms of the mandate, "calling in question our [British] moral right to hold it....

"This restriction of immigration," he said, "within arbitrary limits, unrelated to the economic absorptive capacity of the country, and the undertaking to make its continuance dependent on Arab sufferance, this restriction of Jewish immigration, without any corresponding restriction on Arab immigration, thus having swept away the obligations imposed by the mandate to facilitate Jewish immigration, introducing into the immigration policy, contrary to the specific terms of the mandate, an element of discrimination against the Jews on grounds of race and religion, the reduction of the Jews to the status of a permanent minority—all these things . . . are all grave departures from the terms of the mandate, and they call in question our moral right to continue to hold it. They are not matters within the sole jurisdiction and responsibility of His Majesty's government or even of Parliament, but require the most careful study and examination at the hands of the Mandates Commission."

He warned that the good name of Britain would be tainted if Parliament accepted the White Paper and endorsed it "before obtaining the impartial judgement of the Hague Court."

"This policy," added Herbert Morrison, Secretary for Home Affairs, in joining the debate, "will do us no good in the United States where we need to be doing good and where we need the good will of the great American people." Sir John Haslam recalled a lecture he delivered in Rumania in which he had said: "When Britain has come to a definite decision she sticks to it. Her word is her bond. When she has pledged herself she honors that pledge to the last man and to the last shell . . . ." He then asked: "Can I go back to Rumania and repeat that after the White Paper?"

Lieutenant-Commander Fletcher declared in Parliament: "I think what we have seen in the White Paper is another instance of how the government get out of their difficulties by sacrificing the easiest victims. The government are now joining in the hunt of the Jews which is going on in Europe. Last year, to get out of a difficulty, they did not hesitate to sell the Czechs down the river. This year we see them prepared to sell the Jews down the river."

Left and right condemned the document's "illegality" and "immorality" when it was introduced in parliament in May, 1939. Ramsay MacDonald's son Malcolm, who acted as handy man for Neville Chamberlain and the Colonial Office in the matter, stood pat. "I should have had more respect for him and his speech," exclaimed Home Secretary, Herbert Morrison, "if he had frankly admitted that the Jews were to be sacrificed to the incompetence of the government in the matter, to be sacrificed to its inability to govern, to be sacrificed to its apparent fear, if not indeed, its sympathy with violence and these [Arab] methods of murder and assassination—*that the Jews must be sacrificed to the government's preoccupation with exclusively imperialist rather than human considerations.*"

It was well known that the majority of the House of Commons was hostile to the White Paper. Chamberlain knew that in normal circumstances Parliament would never have approved his new policy. He therefore, at the suggestion of the Colonial Office, which was headed at the time by Malcolm MacDonald, took advantage of the social political atmosphere which prevailed in May, 1939. At such a crucial moment, when the question of war and peace hung in the balance, he felt that the Conservative majority would not dare to embarrass the Government on such a second-rate question as the Palestine one. In this surmise he was proved correct. The Chamberlain government put through the White Paper.

It was not said, and it was never said, that the Zionists had gone far enough in the building of the Jewish national home. It was not said, and it was never said, that the industrialization of Palestine was forging ahead with leaps and bounds and that the Jews were bringing the West to the East and that it was precisely this phase and characteristic of the development of Palestine to which the die-hard imperialists objected fundamentally. Palestine proved in the war that it was the foremost industrial country in the Near and Middle East, way ahead of Turkey, Egypt, or any of the Arab and Moslem countries. It contributed in not a small measure to the success of British arms.

Yet for this very reason, *for being progressive, for being an industrial asset to the British Empire without peer in that part of the world for that reason and no other Jewish Palestine must now die*. For no other reason does the Colonial Office insist that the White Paper of 1939 be made to stand, Jewish immigration be curtailed and a program of intensified agriculture be substituted for the magnificent industrial evolution. The Jews may pride themselves on building a bridge from West to East. The Colonial Office does not want that bridge.

If Malcolm MacDonald and the antiquarian politicians of the Colonial Office would speak honestly and candidly to the Jews, instead of having their diplomatic puppets and journalistic agents becloud the issue before the world with false representations and half-truths, they would say:

"What do you think you Jews are doing in Palestine, modernizing the country, industrializing the land? *You* may think that a splendid achievement, but *we* don't. For with your industrialization you are creating a new working class, also among the Arabs. That working class will in time to come, as has been the experience with every working class in history, ask for more schools, more art, more education, a larger share in culture, a share in the political field, a voice in the running of affairs. If this industrial development in Palestine goes on, the Arab masses will inevitably flock to your side because their economic and material interests are identical with those of the Jewish masses. They will in time come to make a common front with you against their own feudal landlords and against any foreign imperialism which holds the whip hand in the Near East. We do not intend this to happen. We are going to protect ourselves and our own interests. We will not allow your country of Palestine to compete with the British home industry and the industries in countries for whom Britain in normal times acts as commission agent for the Near East. Rather than permit that contingency to arise, we will stabilize and 'freeze' your Jewish industrial development at the present level, bar the entry of new workers into the country, hand Palestine over to the lords of feudal agricultural Arabia . . . .

"We will turn the endeavors of the existing Jewish population exclusively in the direction of agriculture. There is no harm in agriculture. Agriculturally, Palestine is not a menace to our home industry. We do not want any bridges from West to East. We do not want the future. We want to maintain the *status quo*. Think of what your example might do not only in Palestine but in Egypt, in Iraq, in Persia, and in India —in all those huge domains where we are now the masters and the beneficiaries, if we allowed you Jews to continue there would be native industries growing up everywhere.

"You say it is not Socialism you preach and practice but capitalism and that capitalism brings culture, well-being, and a higher standard of living. That is true enough. But capitalism carries Socialism in its bosom. The industrialization of India would create a native working class, tens of millions strong, an army of workers. Among these workers there would inevitably grow up a new pride, a new nationalist spirit, a desire for independence, a constantly larger share in the profits derived from the industrial exploitation of their own land. In the end they would challenge the foreign colonist and exploiter, with the cry 'India for the Indians' and would fight him and expel him.

"To permit industrialization of the colonial areas, beginning with Palestine and extending over the whole colonial world, is tantamount to economic suicide for British industry. Suicide is not a national program. We are not going to preside over the liquidation of the British Empire. In fact, we can wait no longer. War showed us what you have done, that you have gone far enough, much too far with industrialization. No, we must and will check you now. If you resist, if you oppose the White Paper's application, we will fight back. Or rather, we will have the Arab princes fight for us. The emirs and effendis do not want the future either. They also want to preserve the *status quo*. They want to maintain their present privileged position in feudal Arab society. They are our allies. They will know how to present the matter to their poor serfs and to the masses of starvelings on their estates. Fortunately for us, you Jews have not educated the Arab peasants far enough so that they know where the real interests lie. They can still easily be swayed. They can easily be bamboozled by propaganda. We only have to say that you are about to push them into the sea, that new hordes of Jewish immigrants are on the way to rob them of their lands. They still listen to their leaders, whom we will promise—oh, I don't know, an Arab federation, the wealth of Palestine, an empire, the man in the moon, anything . . . . But mark you, if you don't abide by the terms of the White Paper, if you resist, there will be bloodshed in Palestine! The Arabs will attack you! We have consulted with the princes and the big landowners. They are on our side, as the maharajahs of India are on our side in preventing the modernization of that country.

"You Jews had better give up your dreams of a Jewish

state. Accept the White Paper! Forget about the Jewish masses in Europe . . . . We will take care of them after the war. They won't present as big a problem as you seem to think, anyway. Hitler is killing them off at the rate of five or six thousand a day. Three million are slain now. How many can be left in a year or two years? But don't forget, if you resist or if you start raising protests in the United States and elsewhere, there will be trouble. The Arabs can be swayed as on a cord. We have that rope ready. We British won't be blamed. You Jews will be blamed if there is bloodshed. We won't attack you. The Arabs will do that for us and when they do, it will prove to the world that we are right and that you are wrong, that we have done the best we could but that you simply cannot get along together, you Jews and Arabs . . . ."

Of course, they don't talk that way in the Colonial Office. That would be too crude, too cynical, too dishonest. No, they don't talk that way, but they act that way. And actions speak louder than words!

At the height of the battle of Africa, when the sum total of Palestine Jewry's twenty-five years' patient and solid achievement threatened to force itself upon the world's attention in a concrete manner, in that immensely valuable contribution to the British Empire's defense, the Colonial Office quickly set about to mobilize Arab opposition and render it articulate. The invitation to the dance of terror to come was given to the Arab press in Palestine when the prohibition to discuss the White Paper, a wartime measure instituted to preclude the injection of controversial issues into deliberations on the war effort, was suddenly lifted. Lest recognition of the magnificent work of the Jews, their show of loyalty to and their self-sacrifice in Britain's interests, make one clean sweep of all the petty and sordid objections to their endeavor, and at the same time refute the long campaign of obstructionism and chicanery on the part of the Colonial Office as having been an unwarranted and hypocritical policy, proceeding from a spirit of imperialist egotism and racial prejudice, the princes of other Arab countries were with precipitate haste dragged from the obscurity where Montgomery's victory had thrust them.

At a moment when they least expected it, the feudal princes were invited to Cairo to have their say about Zionism and the future of Jewish Palestine. The Jews might think that by their heroic performance they had at last and finally vindicated their position in the Near East. The Colonial Office would show them and the world that they had counted without the Arabs.

The strings were pulled, and the puppets danced to perfection. "Palestine," the Iraq Legation in Washington informed the American press, "rankles in the mind of every Arab." The Iraq diplomats, who are the Colonial Office's appointees, did not say that Iraq is separated from Palestine by the worst desert on earth and that the two countries have no frontier or even a railway line or anything else in common and that ninety per cent of the Iraqi peasants and nomads have never heard the name of Palestine. The squabbles with Britain in the past were blandly attributed by the Iraq legation to "dogged Arab opposition to the British enforced settlement of the Jews in Palestine"! That and nothing else was at the bottom of Iraq's bloody revolt of 1941 under Rashid Ali. "So long as the problem of Palestine was not solved," America was warned, "the Axis broadcasters of Berlin, Bari, and Rome together with Axis undercover men in the Near East and northern Africa will have a powerful emotional weapon against the democracies."

Touching as was this sudden solicitude of the Iraq government for the cause of democracy and the lives of American soldiers in Africa, it proved superfluous, even when the French language newspaper of Henri de Kerillis in New York picked up the torch and handed it on. The Arab masses fell around the necks of the American and British soldiers who came to liberate them from Axis domination. Not a word was said about Palestine and the Jews in Cyrenaica and Tripoli. The starving masses of Libya and Tunisia had other things to worry about than Levantine backstairs politics.

But that does not matter. Any stick is welcome to beat a dog. When the new, reinvigorated anti-Zionist policy of the Colonial Office is to be enforced in full and justified, we are surely going to hear of the claims, demands, and views on the matter of the Algerians, along with the views of still more distant Arabs. De Kerillis can be quoted in person. One would, as a matter of fact, not be surprised to see the Moslems of Java brought forward to have their say in days to come or the Achinese of Sumatra, who are of pure Arab descent and who were the only inhabitants of Indonesia who sided openly with the Japanese against the Dutch. Is it not an established policy since Darlan to consult one's enemies and to slap one's most loyal friends in the face? Of this the Messrs de Gaulle and Chiang kai-Chek could perhaps testify!

Even so, Iraq had come into the war at last—and on our side! For this we were unquestionably to be deeply grateful, although it would take some time, no doubt, before the camel rustlers of the Nejd border were transformed into those tough legions that will accompany the Anglo-Russian allies in the assault in *Festung-Europa*. But why did Iraq come into the war? Was it to put an end to the Axis regime of slavery, racial hatred, and blood? Was it to send at least a token force, belatedly, to the assistance of Britain and thus to a certain extent redeem for Baghdad's earlier betrayal?

No, not that! Iraq came into the war, stated its spokesmen in Washington, only for one thing, to settle one question, and one question only.

And what could that question be? What claims was the kingdom of Iraq, which had to be beaten into line, going to make in the international councils? What difficulty was still barring the way to fulfillment with these traditionally incurious sons of Ishmael who but yesterday, so to speak, had finished a term of five hundred years under the Turkish harrow? What was the problem uppermost in their minds? Was it the old dispute over the grazing rights of the Bedouin on the Nejd border? Or the extreme poverty of the urban proletariat in Baghdad and Basra, the disease and squalor rampant in the former Garden of Eden, or the almost universal illiteracy of Iraq's inhabitants? Were these some of the vexations preying on the Iraq Government's strange and dark imaginations?

No, nothing as trivial as all that. The Iraq government came into the war to "have a seat at the peace conference and to have a say in the affairs of . . . Palestine," a country that does not threaten anybody in the whole world and least of all Iraq, where every man cultivates his own garden and which incidentally did a lot more useful work in the Allied cause than staging revolts.

Not Eretz Israel, the land of Israel, which had strained its every nerve and energy to the breaking point to make England victorious, but Iraq, the country that had called in Hitler's Luftwaffe, was suddenly given a place in the ranks of the United Nations. Not the Palestinian Jews, who had offered their all, were called Britain's "noble ally," but Ibn Sa'ud, who could not spare one man, one mule, or one camel when the Eighth Army stood panting for breath with its back to the Nile Valley. Palestine was Eretz Israel (Land of Israel) no longer—that official title was withdrawn and prohibited by the Colonial Office—but Abdullah, the Emir of Trans-Jordan, was allowed to talk in the British-kept press of Egypt about his new throne in Jerusalem, the capital of Eretz Israel. It is true that Ibn Sa'ud, who had delegates at the preliminary conference called by the Colonial Office in Egypt, did not say he would enter an Arab federation, but he did say that he opposed the Zionist settlement of Palestine. And that made it unanimous, for the King of Egypt was also in agreement on that point . . . .

It was done, the great transaction! The gentlemen of the Colonial Office need worry no longer to reformulate their own refuted and discredited objections to the Jewish national undertaking in the Holy Land. That had become superfluous. The Arabs were doing it for them. The Arabs and others!

Sir Ronald Storrs, the former Governor of Jerusalem, was suddenly recalled from Rhodesia by the Colonial Office. He made a quick tour of the Holy Land and after a week's sojourn uttered a solemn warning: "The Arabs are restless. I fear that there will be grave trouble in Palestine if the Jews do not quickly abandon their opposition to the White Paper." Sir Ronald is an expert, an impartial expert—hasn't he just come from Rhodesia? But he always had a fine flair for detecting trouble in the wind. "They are arming," he said. "Both sides are secretly collecting arms!"

Both sides? And what was the Palestine administration doing? The Palestine administration, which is a branch of the Colonial Office, imports known terrorists, Sir Ronald! In India it puts Gandhi, the apostle of nonviolence, in prison. In Palestine it sets loose advocates of murder, arson, and rape. The Palestine administration has brought back from jail or from the Seychelles Islands, where they were for their share in the wave of murder and terrorism in 1936-1939: Jemal Husseini, cousin to the fugitive Mufti of Jerusalem, member of the Arab Higher Committee; Auni Bey Abdul Hadi, who was implicated as a leader of terroristic activities, and Hussein Fakri el Khaldi. At the same time, it permitted to return to Palestine close friends, relatives, and supporters of the Mufti who had been arrested in Persia following their complicity in the Iraq uprising, among them such well-known figures as Musa el Alami, Amin Tamini, and others. It released from custody murderers "for good behavior" . . . in jail and gives them liberty to start their nefarious agitation all over again.

Yes, there is going to be trouble in Palestine. The trouble was that there was no trouble. That is why there is going to be trouble. The wave of terror that swept over the country from 1936-1939 abated the moment war was declared and the Axis agents and instigators departed at last. For three years the Palestine administration refused to molest these Nazi emissaries who were the Mufti's allies. They went about the Jewish land flying the swastika on their motorcars—the swastika in the land of the people whom Hitler had begun to exterminate. They egged on the Arab peasants. In several instances they participated in terroristic depredations against Jewish persons and Jewish property. But when they left, Palestine grew quiet at once. Palestine went to work. The Arabs started to earn money. They made money hand over fist mainly in Jewish enterprises. Jewish moving-picture theaters, dance halls, cafés in Jerusalem, and beaches in Tel Aviv and Haifa were crowded with Arab clients. They fraternized with the Jews. They worked with them. Their children started to go to Jewish schools, Jewish technical colleges, model farms, playgrounds, clinics. They held big demonstrations together, Jews and Arabs. They revealed (what was already known to the Palestine administration) that during the terror of 1936-1939 more Arabs had been slain by Arabs than Jews by Arabs. Arabs confessed that they had been intimidated by a small, terroristic minority in 1936-1939. In many colonies they had come to ask forgiveness from the Jews for their past behavior. They spoke in admiration of the Jewish national policy of *havlaga,* self-restraint, the policy of non-retaliation, of not returning evil for evil, which had been practiced and strictly enforced by the Jewish community during the terror. There was no trouble at all, not a trace of it. The past was being forgotten as everybody worked . . . .

And now there is going to be trouble, Sir Ronald says. Who then makes that trouble? The seeds of explosion which have brought Palestine to its present state had been sown long ago. The Palestine administration forbids demonstrations of solidarity between the two peoples. Jews and Arabs may not belong to the same labor unions. They may not go to the same meetings together. They may not hold a picnic together. A Jewish newspaper printed in Arabic and devoted to Arab-Jewish amity is forbidden. The ban on controversial subjects in the press is lifted. Terrorists are brought back. Conferences are called in Cairo, and the news is given out that everything is to change in Palestine: Abdullah is to get a throne in Jerusalem, Palestine is to be incorporated in Syria, Ibn Sa'ud is to take over! Of course there is going to be trouble when trouble is made!

All at once the correspondents, the diplomatic, military, and political observers who guarded an unbroken silence on the subject of Jewish fighting qualities, on Palestine's amazing contribution to the victory in Africa, find their tongues again. Newspapers, magazines, and periodicals in America are flooded with articles, dispatches, and editorials warning of the tension in Palestine and of a coming explosion. Sir Ronald Storrs is quoted right and left, and with him the British agency in Cairo, Arab politicians in Taif and in Baghdad are all made to have their say. Trouble coming! Trouble coming!

The whole Near East—Jewish Palestine especially—is presented to the American public's attention now as a cesspool of intrigue and deceit, then as a dangerously exposed powder barrel which may explode any moment. For long months efforts have been under way to prepare the American people psychologically for disastrous news from the Holy Land. Anything, they are made to believe, may happen there, any time. But when it does happen, which may well be before these lines appear in print (for the Colonial Office is in a desperate hurry to make it happen soon), the responsibility for it must be fixed. And upon whom shall it be fixed?

One day, we learn from a Near Eastern correspondent, after he makes a call on Ibn Sa'uds Foreign Minister, that political Zionism is a constant provocation to the Arabs. The next it is whispered about in Washington that seven American divisions have had to be diverted from the fighting fronts to watch the situation in the Near East, where Arab patience with Jewish encroachments upon their rights is nearing exhaustion. Did we know, asks one publicist, that the Jews were smuggling arms into Palestine? Did the Jews perchance enlist so enthusiastically in the British army, suggests, with criminal naïveté in a Jew, the correspondent of the anti-Zionist *New York Times,* in order that they may familiarize themselves with the manipulation of modern mechanical arms and battle tactics.

Did we know, he asks, that Jews were intimidated by their own extremists into joining the British forces?

One hundred and thirty-seven thousand Palestinian Jews voluntarily registered for military service within a month after war was declared. The British Information Service in New York issued a statement on July 16, 1943, to explain that Britain is forced to curtail enlistments in Palestine because they threaten to deprive the Palestinian economy of indispensable mechanical experts. But *The New York Times* knows better; it impugns the motives of its owner's fellow Jews in their fight against their butcher and makes of the valuable contribution of the Palestinians a piece of vile intrigue.

"Seriously concerned," as more than once it declared itself to be, "with the safety and security of the country," the Palestine administration gave the direct lie to that declaration when it launched a public trial of two or three British soldiers accused of implication in a gun-smuggling plot. An administration, the most moderately qualified for its duties, would know that such things are not discussed in public in a time of tension . . . unless that administration itself desires to see the tension heightened and the pent up steam brought to the point of explosion. The public prosecutor, who pleaded for leniency in behalf of the accused on the ground that they had been caught in the sinister meshes of a powerful gunrunning racket he sought to link with the Jewish agency, exclaimed that if the charges were true "the whole world ought to hear" that the agency had been "working to the detriment of the Allied war effort."

After having been kept in total ignorance by the

same Palestine administration and by all other British agencies and information services on the subject of the extraordinarily significant and important war effort of the Jewish community of Palestine, the world was suddenly to be impressed with the vicious lie that the Palestine Jews are conspirators, gunrunning racketeers, and seneschals of sedition and that the Jewish agency, which is the governing body of the Jewish national and international reconstruction effort, is some mysterious, sinister sanhedrin of conspiring Elders of Zion . . . while Arabs' apprehensions are aroused by the threat to their lives and property implied in the accusation that the Jews are arming themselves.

David ben Gorion, one of the directors of the Jewish agency's executive in Jerusalem, whose only son died in the defense of Tobruk, denounced the trial—rightfully, I think—as "characteristic of the lowest type of anti-Semitism" and as "a crude frame-up designed to defame the Jewish people." And surely; as a piece of work of the *agent-provocateur* that trial will rank with Göring's firing of the Reichstag and with Father Gapon's march into the Winter Palace square in St. Petersburg.

Thus the web is being woven around the Jewish people for the day when the Colonial Office finally puts through its design to bring the building of the Jewish National Home in Palestine to a standstill. It cannot be done by legal means, so it is done by means of vicious intrigue and provocation.

Are Palestine's Jews secretly arming? I don't know. But would you, reader, not get yourself a gun if you saw that avowed and convicted terrorists are set free to maraud around your home? Were the American colonists wrong when they made powder and bullets in secret to be ready for the day when their besotted king would enforce his unjust decrees?

These men the Colonial Office calls to Cairo—Ibn Sa'ud's representative, Abdullah of Trans-Jordan, Nuri Pasha of Iraq, Nachas Pasha of Egypt—are as much Britain's puppets as Antonescu, Mussert, Quisling, Mannerheim, and Horthy are Hitler's. They do what they are told to do, especially after the defeat of Rommel blasted their hopes of throwing off British suzerainty. They are marionettes, without a shred of independence, who dance to the tune that is played in the British agency in Cairo.

The Arab terrorist faction has not been welcomed back by the Colonial Office because the Jews were said to be arming. If the Jews are arming they do so because the terrorist faction has been revived and because they remember its deeds of arson, rapine, and murder in 1921-1929, and 1936.

In order that there shall be no sympathy for Zionism on the day when Britain clamps down on the building of Palestine, some American diplomats, strange to say, have joined the imperialist British conspiracy. Their job is to prevent an outburst of indignation on the part of American Jewry, the Free Church people in the United States, and labor when the whole ugly maneuvre of the Colonial Office come to the light of day. The Jews in America are put on their good behavior. Do not say anything about the White Paper! Do not project Palestine's political problems on the American scene! You will prejudice the war effort! You are disturbing Anglo-American unity! Wait till after the war! Palestine is dynamite! It is going to explode! Leave the White Paper alone!

Some liberals, confused by all the fuss and the threat of trouble, want to speak a calming word: why not let that White Paper stand? Think of the "poor Arab"! The poor Arabs who own ten half-empty empires, from the Persian Gulf to the Straits of Gibraltar, and whose kinsmen in Palestine are the best-looked after Arabs in the world . . . .

Poor Arabs! Poor Palestine! All the old arguments are trotted out: Palestine can never take care of two million Jews. If that is true, how can it take care of millions of Arabs? Are the Arabs in trouble? Is Hitler exterminating Arabs? Are the Arabs a homeless people? Have the Arabs been driven from pillar to post these two thousand years? What do the foreign Arab princes want with that narrow strip of coast on the Mediterranean that is not even mentioned by name on the average map when they have Iraq (Iran), Trans-Jordan, Syria, Lebanon, Yemen, the Hadhramaut, Saudi Arabia, the Nejd, Egypt, the Sudan, Tripoli, Tunisia, Algiers and Morocco?

There is going to be trouble in Palestine! Of course there is! Lord Cranborne told the House of Commons: the Arabs fear that the Jews will crowd them out, push them into the sea, swamp them, and rob them of their land.

Lord Moyne exclaimed: "This much is certain: the Arabs are never going to leave the bones of their ancestors in Jewish hands!"

That was a new argument of which nobody had ever

thought before: the bones of ancestors—another source of trouble!

When the Archbishop of Canterbury in the House of Lords adjured the British Government that in the matter of Palestine Britain stands before the bar of God, history, and humanity and asked if nothing then, really nothing could be done about bringing Jews out of the Nazi hell into the Holy Land, Lord Cranborne first reprimanded him and then said: "The most reverend Primate forgets there is a grave political problem in Palestine!"

Can anything more hypocritical and perfidious be imagined? Who created that problem? The Arabs? "The Arabs are an excuse," exclaimed Lord Wedgwood; "they are not a reason!"

The Colonial Office first creates the problem, diligently, never tiring for a quarter century, and goes on today deliberately accentuating it, and then dares to have its spokesmen in Parliament, at the Bermuda Conference, and in Washington bring that problem's existence forward in justification of barring Jewish refugees!

CHAPTER 6

THE SOLUTION

THE real problem that faced King Feisal and all of Iraq's successive administrations since the country was torn loose from the Turkish Empire at the cost of a hundred thousand British lives was and remains the intense poverty of the Iraqi people. This is the rock which has wrecked all attempts to whip the two or three million peasants, nomads, and camel breeders who inhabit and roam* through the Mesopotamian vilayets into some semblance of nationhood. Of what use are rudiments and paraphernalia of modern democratic government, as they exist in Baghdad—a parliament, foreign diplomatic representation, governmental departments, newspapers, and an intelligent bureaucracy—if the mass of the people are strangers to it all and derive no benefits from it, but continue to wallow in abysmal ignorance, starving half the time, afflicted with a destitution the depth of which can scarcely be imagined in America? Baghdad, where a small superior caste of landlords, merchants, politicians, professional soldiers, advocates, and intellectuals and their journalistic hangers-on indulge in the luxury of national grandeur, is a sounding brass and a tinkling

*There are no precise population figures on Iraq's Bedouin.

cymbal, full of fury and sound at times, but without anything substantial at its back.

Baghdad has never been able to provide the eternally wandering and raiding tribesmen in the endless hinterlands with a banner or cause around which to rally and, to use a homely expression, to stay put. And those tribesmen, it must be owned, could not have acted otherwise than they did. They had to go raiding and had to keep on moving: hunger drove them. There was not a single common aspiration strong enough to bind them and hold them together, to set them to work, and to keep them occupied with one collective task, the execution of which, or the mere desire to execute which, would in course of time perhaps have engendered in them a sentiment of common destiny. Instead, they were by a dismal necessity kept at dagger's point among themselves and with all their neighbors, Arabs like themselves.

This was the all-overshadowing problem that confronted Feisal. The Iraqi leaders must often have despaired of producing the miracle of transforming the tribesmen into one single-minded, one single-aim-pursuing people. I know King Feisal did, for he told me so more than once, the last time in Geneva, shortly before his untimely death. And if Feisal was in a quandary, how much more must his successors—men of lesser genius and organizational ability—have looked upon the dream of creating a national consciousness out of the desert's sterile heat as a well-nigh impossible task.

Everything was tried, one might say, in the course of the quarter century that elapsed since the Ottoman

Turk went home to Anatolia, every expedient to weld the tribesmen into some cohesive shape of nationality. More than once the promise of loot and plunder was held out as the strongest inducement to prick and activate the dormant national faith. Several times the Bedouin were roused to a holy war against the British, the hate of the stranger in our midst having at all times been the most obvious and useful instrument to rouse the primitive chauvinist spirit. But the British struck back and they struck back hard. Then came the Assyrian massacre in 1933, and the flame of Iraqi national passion flared high for one ghastly moment only to peter out when most of these Christians, members of one of the oldest branches of the Church Universal, had been killed and their spiritual chief, the Mar Shimum and his chapter, had been rescued by British troops from the pursuing warriors. Once more the Iraqi tried it: under their leader Rashid Ali in 1941. Then they were beaten into impotence, and their leaders fled to their paymaster in Berlin who lodged them, with significant irony, in the house that once belonged to the Jewish statesman, Walter Rathenau.

"Arabs," said Lawrence, who knew them, "can be swung on an idea as on a cord, but they escape the bond when success has come, and with it responsibility, duty, and engagements." When the idea is gone, the work falls in ruins. No idea to kindle the national consciousness has been produced in the Arab world that the waves of time did not quickly wash away.

Until there appeared on the scene Nuri Said Pasha, who was elevated to the premiership of Iraq when the British army, having defeated the Iraqi, fought its way into Baghdad for the third or fourth time in a quarter century. Nuri Pasha was Britain's man, the candidate of the Colonial Office. He is not altogether a novice in international Arab affairs. He was one of the politicians who established contact with Feisal and Lawrence at the time they were preaching the revolt against Turkey. He is an old comrade of British military men and a frequent caller at the agency in Cairo. Since those heroic days of the trek to Damascus he has been under the influence of that school of British colonial administrators which pretends to see in an Arab federation under British aegis the best solution of the problem of the Levant's political chaos. This makes him *ipso facto* an opponent of French influence in Syria and of the transformation of Palestine into a national home for the Jewish people. His chief advisor in Baghdad is one of the oldest and most influential functionaries of the Colonial Office in the Near East, Sir Kinahan Cornwallis, His Britannic Majesty's ambassador to Iraq, who was another of Lawrence's companions on the road to Damascus.

Whatever policy Nuri Pasha formulates, therefore, comes at the suggestion of the Colonial Office, and this includes one of his first official acts in the premiership, which consisted in refusing transit visas to a thousand Jewish children who had been saved from the pogroms in Poland and who were marooned in the Soviet Union and Persia for lack of an Iraqi *laisser-passer.* Not in covered trucks, as was suggested, not by night, not even in airplanes, "never and under no circumstances will these Jewish immigrants pass through or over Iraq territory," said Nuri Pasha who had just in the name of his country subscribed to the humanitarian ideals of the Atlantic Charter.

The refusal to let these little children pass served a twofold end: it showed the independence of mind of the new Prime Minister of Iraq and at the same time gave British colonial officials in the Near East added cause to impress the Palestinian Jews with Nuri Pasha's intransigent hostility: it is not we who are hostile to Palestine, you see it for yourselves now. We are impartial, but you Jews had better keep the Arabs of Iraq in mind.

Confronted with the same problem that harassed every successive administration in Iraq since the country's liberation from Turkey: to find a unifying principle, a policy, a creed that by its simple inspiration would rally the mass of the peasants and nomads around him for the performance of a collective task of national regeneration, Nuri Pasha does not, apparently, propose to act differently than any of his unsuccessful predecessors. Instead of coming forward with a program of social and economic reform pertinent to the backward social and economic conditions of his country, he is about to launch into a chauvinistic foreign adventure in order to avoid facing and resolving the flagrant contradictions of class and caste in Iraqian society itself.

It is the old trick of the European reactions: when you do not see a way out of the difficulties at home, invent a diversionary maneuver, best of all plunge into a bloody foreign enterprise. And the first victim is nearly always, and certainly in this case, the Jew.

The functionaries of the Colonial Office in the Near East think to have found in Nuri Pasha the whipping boy to destroy by violence in Palestine what could not be stayed by their own schemes of niggling dishonesty and provocative inconsistency.

Nuri Pasha, the agent of the Colonial Office, speaks to his tribesmen of something that does not concern them in the remotest way: Palestine and the Jewish national home. He rouses their envy, their cupidity, and their hungry and destitute man's instinct for loot. Not a month passes without the Colonial Administration arranging for delegations of Iraqis to visit Palestine and to be shown around Jewish farms, Jewish plantations, and Jewish industrial and cultural installations. These delegations see Tel Aviv, the Paris of the Levant, with its fair, its cafés, theaters, banks, modern shops, and well-dressed, clean people. They see Haifa, expanding industrially; its steel mills, cement factories, glass plants, silk spinneries, linen weaveries, and laboratories; Jerusalem spreading out gloriously over the neighboring hills, its colleges, broad new boulevards, university, libraries, hospitals, the offices of the great co-operatives; the valleys of Esdraelon and of Sharon teeming with activity, with their hydro-electric stations, irrigation works, dairy farms, endless orange groves.

And when then they have seen it all, they go home to Iraq, to the waste places of Mesopotamia, to the black desolation and unalloyed mass destitution of their homeland, not inspired to go and do likewise, but sullen, bitter, disillusioned, burning with envy. And then comes Nuri Pasha, the British agent, after attending one of the so-called Pan-Arab conferences at the British Agency in Cairo, and says: "But that Palestine you just saw, didn't you know? It really belongs to us!

It is ours, yours and mine, for the asking. Our ally, Great Britain, favors the establishment of a Pan-Arab confederacy. That federation is to include the Holy Land, and then where will the Jews be? You worry about the Jews? They will be a very small minority in a great Arab Empire, I assure you!"

And what do Arabs do with minorities? What did they do with the Christian minorities in Lebanon, in Syria? Do you remember what happened to the peaceful Assyrian minority in Iraq in 1933 when the same Nuri Pasha was Minister of the Interior? They were simply massacred, and there was no interference. The civilized world, the Christian world, stood aside as usual and excused itself: "One must not and one cannot interfere in the purely internal affairs of a sovereign state like Iraq. That would not be proper or legal!"

"The Jewish Homeland in Palestine," announced the Iraqian legation in Washington "rankles in the heart of every Arab." It does not rankle at all. It is being made to rankle by Nuri Pasha and his masters, the gentlemen of the Colonial Office!

Palestine's condition and situation are of no concern, either directly or indirectly, to the Iraqi tribesmen. Nuri Pasha has no more legal right to interfere in the affairs of Palestine or decide on that country's future than he has a right to speak of what is to be done in Ireland or Siam! Palestine is the business of the Permanent Mandates Commission of the League of Nations, of The Hague Court, of the government of the United States of America which ratified the British mandate over the Holy Land on the understanding that Britain fulfill her pledges to the Jewish people. In the final analysis it is the business of the Jewish people of the world for whom it was set aside by international sanction. Yet today the future of Palestine is being placed in the hands of Iraq's peasants and nomads. They are being talked into and being whipped up into making Palestine their business, so that they will forget, for the time being, their real and just grievances against their own landed aristocracy, which keeps them in a condition of abject poverty and subjection.

That is Nuri Pasha's policy, and the policy of the Colonial Office. That is the rope idea on which the Arabs are to be swung in the performance of their next great historical task: crush the Jewish National Home, stunt its growth by violence, check its expansion, ruin its prosperity, kill its hope . . . .

Nuri Pasha is a man of lucent inspiration. His Washington legation did not call him "astute" for nothing. He has found the slogan, the issue, the banner around which an inchoate, illiterate, poverty-stricken swarm of Bedouin can be rallied. That slogan is: the Jews! And a valuable slogan it is! Didn't Herr Hitler manage to whip up a nation of *Denker und Dichter* into paroxysms of frenzy with that same watchword? Wasn't the German youth hurled into an infernal holocaust on the simple bloody motto: "Perish Judea!" What Herr Hitler did with the materially most advanced nation in Europe Nuri Pasha can surely do with a mob of traditionally misled runners-amok!

Can he? Are the United Nations, after winning a victory over Iraq, to permit that country's Prime Minister to use his nation's defeat as an instrument to extort from us a benevolent attitude towards the basic, anti-

Jewish tenets of the Nazi philosophy which Nuri Pasha brazenly announces as the foundation of his country's future policy? In order to soften the pangs of her defeat, is Iraq to be allowed a looting expedition in the direction of Jewish Palestine? Why do the British and American armies fight anyway, and why did the Jews of Palestine under Raziel save Iraq's oil installations from destruction by the Nazi marauders and their Iraqi henchmen? Do we not care for the peace so long as we win battles? Is it for the defense of human rights and dignity and national freedom, also for the Jewish people, that humanity goes through its present agony, or is all well the moment the pipe lines in Iraq or elsewhere are reasonably safe? Was Anatole France right after all when he said: *"Men think they die for their country, in reality they die for big business."*

After Rashid's flight and the dispersal of the last rebellious bands, Nuri Pasha stepped forward to lead the country "in the path of peaceful collaboration with Great Britain." He was known to be Britain's man. For this reason he was also one of the most unpopular figures in the country, and his new position as Prime Minister was extremely shaky. Other predecessors than Rashid in the premiership of Iraq have had to pay with their lives for their pro-British policies. It is true that Nuri had at his back the British forces who had just crushed the rebellion led by his predecessor. But among the people of Iraq he found no support. His accession to power did not come about as the result of popular elections: they do not exist in Iraq. He was appointed by the Regent, who could scarcely do otherwise in the presence of a victorious British army than install a pro-British politician in the premiership.

From this point of view Nuri Pasha's acceptance of the office must be considered an act of high courage indeed. It took nerve to announce a complete reorientation of Iraq's foreign policy in the teeth of the smoldering anti-British sentiment in the country and the disillusionment of the army officers who still smarted under the defeat administered by the British forces. The Iraquian legation in Washington announcing his appointment called Nuri Pasha "the most popular man in the country." This was a little trick of Oriental fantasy, if not clever Anglo-Saxon-inspired propaganda directed at the Americans who think it natural and proper for the highest office to be filled by the most popular person. But his popularity Nuri Pasha still had to win with the people or rather with the landowners' class, the Moslem clergy and the upper circles of the officers' corps.

However, this task did not prove too difficult with the assistance of Sir Kinahan Cornwallis. For it is precisely with those castes in the primitive society of Iraq whose interest it is to keep the kingdom functioning on a predominantly agricultural semifeudal basis that Great Britain prefers to deal and with which she has proven that she can deal best. For the interests of those classes are as identical with those of the colonial power as are the interests of the princes of India with those of the British Raj. They are inseparable. Only under the protection of a modern colonial power can the landed aristocracy of Iraq be assured of maintaining undisturbed its position of pre-eminence in society. Nuri Pasha's task consisted in bringing this truth home to the effendis and muftis and the scions of the noble families who occupy the chief positions in the army of Iraq.

On the other hand, the interests of the masses of Iraqis are not identical with those of the colonial power, any more than they are identical with those of their own ruling class. As a matter of fact, with both they are diametrically at variance. Only the masses are not aware of this circumstance. Iraq's masses, like the masses of India, have not learned yet to present their wants and needs with decision. Iraq's depressed and demoralized peasantry—demoralized by poverty—has not yet learned to attach their indignation to the right object and to remedy the ills from which it suffers. Even the latest defeat of Iraq's army has not brought the lesson of history home to them. They still leave their affairs and their lot in the hands of their feudal overlords, who are now, under Nuri Pasha, holding before their eyes the mirage of a Pan-Arab Empire, the establishment of which Great Britain will never, so long as she is able to prevent it, allow to materialize in the neighborhood of her own most vital interempire lines of communication, where she has just defeated the Italians and Germans, who tried to install themselves, and has eliminated the French, who had a foothold there . . . .

Instead of chasing an evanescent phantasmagoria, the Iraqis' efforts should be directed towards economic emancipation, towards freedom from want through a resolute policy of modernization, industrialization, and irrigation which would give their country back its former marvelous fertility and wealth. They do not realize that, in order to keep them from making such demands and from striving in that direction, their aristocracy and their political chiefs, under inspiration of the Colonial Office, will seek again and again to divert their attention from what should be the major issue in their lives to matters of such relatively small intrinsic importance to them as the Judeo-Arab question in Palestine or to foolish pipe dreams of Arab imperialism.

The Arab overlords, assisted by the Colonial Office, are out to prevent reason and right from taking their course in the social conflict in the Near East by imitating the precise method that Hitler and the Czar of Russia before them used in setting up the Jew as a scapegoat.

This may not be Britain's intention. I cannot imagine it to be Britain's intention. It surely cannot be the desire of the British people who are possessed, perhaps in a larger measure than any other people on earth, of a high sense of justice and fair play. Nevertheless, in practice, through that curious divergence of British popular feeling and the inhuman *Realpolitik* of a handful of imperialistic permanent officials in the Colonial Office and in the Palestine administration, in practice, alas! it definitely works out that way.

When I once remarked, in a public gathering in New York,* that the decision of the Colonial Office to bar all further immigration into Palestine, precisely now,—the most harrowingly tragic hour in Israel's long

*In Carnegie Hall, May 2, 1943 on the Day of Compassion for the Jews proclaimed by the Federal Council of Churches of Christ in America.

and melancholy history, leaves Germany in the role of the exterminator of the Jews and Britain of the Power barring their rescue, I was severely taken to task by several British agencies. I was called grossly unfair. I had placed a false emphasis on facts. My motives for speaking were questioned. It was even suggested that I tried to please Herr Hitler and that I had acted prejudicially to the war effort by injecting a note of discord among the United Nations.

Yet, what are the facts? Thousands of the humblest and poorest Jews, inoffensive and innocent men, women, and children, are being murdered each day, smothered to death by poison gas, electrocuted in abattoirs, buried alive, mowed down by machine-gun fire. Newborn infants in sight of their mothers are being kicked about as playballs by German soldiers. A Jewish father is compelled to carry the severed heads of his two children to the place of his own execution through a laughing crowd of Nazis. What is happening to the Jews in those somber regions of eastern Europe surpasses in horror anything imagined by Dante in his vision of hell. Three million Jews have been slain. Yet though the military threat to the Near East has been removed, the doors of Palestine are kept shut, in spite of the fact that agronomical, economic and ethnographic experts from America, France, and Holland declare that Palestine has ample room to take in, integrate, and absorb vast numbers of Jews.

The Jews remain barred from Palestine in spite of the fact that the Evian and Bermuda conferences on refugees have shown conclusively that no other state on earth is prepared or willing to accept Jewish immigrants en masse. The doors of Palestine are not opened, even when it is definitely known that some of Germany's satellite states, with the defeat of the Axis staring them in the face, are inclined to stop the horrible deportations to Poland they carry out at Germany's behest and to let their Jews proceed elsewhere if they can obtain travel facilities. When these are the facts, who then bars the rescue of Jews? Where does the responsibility lie?

I say that it is an injustice which is not diminished in gravity by the greatness of the nation that commits it. I cannot say otherwise. And as to the allegation about trying to please Herr Hitler, I can only say that the *Frankfurter Zeitung* of January 15, 1943,* named me second, after Winston Churchill himself, along with ten other men and women as being chiefly responsible for bringing America into the war on Britain's side. When some seek to daub existing vices with virtue and charge that denunciation of evil amounts to creating disunity, I answer that I will not be deterred from repeating what Churchill, Atlee, Morrison, and Lloyd George once called the "immoral," "illegal," "unjustified," "cowardly," "contrary-to-Britain's-interests" action of the Colonial Office and of the appeasement government of Chamberlain in tricking Parliament into passing the White Paper of 1939, which, if it is now fully enforced, dooms vast numbers of Jewish human beings to nameless suffering and a great people to continued bitter exile, besides branding the fair British name with the odium of betrayal.

---

*Reported in New York by the French-language paper *Pour la Victoire,* issue of Saturday, March 6, 1943.

I will go on saying and writing that this is an injustice, though American Jews themselves, intimidated by England's might and the sophistry of imperialist deductions, remain fearfully silent, though Britain mobilizes newspapers and publicists in America to talk the crooked straight, and though I am slandered as a paid agent of the mythical Elders of Zion or as a non-Jew who has no business to stand by the Jewish national cause. I will go on as long as the injustice goes on. La Fontaine said that the right of the strongest is always the best. But it is not a moral right and in a Christian world it must not prevail. I will denounce it wherever I can, as long as I can, until right is done, until Israel is back in his own house, in Eretz Israel, in the land promised to Abraham by God Himself. With a slight paraphrase on the words of Erasmus, I say: *Angliae amicus sed magis justitiae!**

"The little notch of Palestine," as Balfour called the Holy Land, cannot by itself solve, and cannot be made to solve, the staggering problems of the fifty times larger Arab countries. Whoever holds out a prospect to the contrary to the Arab peoples is guilty of deception and cruelty. He either takes advantage of their political immaturity and their childlike cupidity or he caters to the inordinate personal ambitions of certain Arab politicians and to the rapacity of the Arab feudal nobility.

If the British government is sincerely concerned about the condition of the Arab peoples who are at present, without an exception, within the orbit of the

---

*I am a friend of England, but I am a still greater friend of justice!

British Empire's sway and influence and therefore Britain's wards, it will order the functionaries of the Colonial Office in the Near East to cease pricking Arab jealousy and hunger by holding up the false mirror of an eventual possession of Jewish Palestine as the nostrum for all the Arabian countries' ills. In order to lift the Arab peoples out of the economic, political, and cultural stagnation into which they are sinking ever deeper, more heroic, more radical, and more civilized measures are necessary than turning over Jewish achievements and acquisitions in Palestine to the plunder of foreign marauders. For this would be the ghastly upshot if the sinister political intrigue now pursued by the Colonial Office with regard to Israel's heritage and the fate of the Jewish people were carried through to its ultimate consistency.

Even if the Jewish Land were to be included in a Pan-Arab federation, as the Colonial Office plans, and for which plan it has mobilized Britain's puppets—Ibn Sa'ud, Nuri Pasha, Abdullah, and Nachas Pasha—and for which it surreptitiously seeks support among Washington's diplomats, it would not provide a permanent solution for the woes of Iraq, Egypt, Trans-Jordan, Syria, and the Hejaz.

The surrender of Palestine into the hands of foreign Arabs would perhaps bring a brief surcease in the inarticulate but constantly mounting pressure engendered by the poverty of the foreign Arab masses. There might be a moment of respite, a breathing spell until the loot is dispersed and absorbed after a short and bloody grab and plunder fest. But on the morning after the great anti-Jewish St. Bartholemew, the situation in adjoining Arab countries would remain fundamentally unchanged. Palestine itself would in a quarter century revert to what it was in the days of the Turk: a wilderness of decaying cities and wasting land.

But if Palestine cannot solve Arabia's problems, Iraq can. Iraq can indeed become the chief component of an Arabic federated empire or commonwealth of states. For Iraq is an empire in itself, a land of fabulous fertility and untapped potential richness. Iraq is three times the size of England and Wales combined. But whereas England and Wales have an aggregate population of thirty-five millions, Iraq, by the most generous estimates, has little more than three millions. Of these some seventy per cent are nomads without a fixed place of abode. There is room in Iraq, conservatively speaking, according to the estimates of English experts such as Sir William Willcocks, Dr. Philip Ireland and many others, for a population of fifty millions, forty-seven million more than she has at present. Once empires flourished on Iraq's territory, and a peasant population of thirty millions tilled her soil. Today there is less than one million persons on the land.

In the past the British and Indian governments have extensively and seriously studied the possibility of making Iraq an outlet for millions in the almost inhumanly overpopulated Indian Empire. These schemes, however, were opposed by the Iraqi authorities on the grounds that the influx of these non-Arabs would soon obliterate the Arabic character of their nation and country.

Today the outlook has changed. Jafar Pasha al

Askari, Prime Minister of Iraq in 1926, informed the Royal Central Asian Society: "What Iraq wants above everything else is more population. This is the fundamental and necessary condition of progress. Without more people Iraq is doomed to stagnate," he said, and doomed, we may well add, to chase foolish will-o'-the-wisps in the form of foreign political adventures.

"In the Nile Valley, from Asuan to the sea," the Iraqi statesman continued, "where you have a riverain population living upon irrigated lands, there are some fourteen million inhabitants. The possible irrigable area in Iraq is certainly not less than that of Egypt . . . ."

The British government's *Report on Ten Years' Progress in Iraq,* 1920-1931, speaks of the "gravely insufficient agricultural population," which is "a hindrance to the development of irrigated farming" and adds: "Real agricultural development in Iraq will come through an increase of agricultural population . . . . for land brought within schemes of irrigation extension."

And here is a more recent view (October 27, 1938) of a Middle East expert, H. T. Montague Bell, writing in *The Times,* (London) after long years of study of Iraq's problems: "Iraq's paramount and desperate requirement is an increase of population . . . . Her present inhabitants cannot do justice to the potentialities of the land—the lack of labor is a constant problem—and she is at a disadvantage against Turkey and Iran with their far larger populations. The settlement of the nomads on the land may add to her wealth, *but any substantial increase of population in the near future must come from the outside."*

Somewhat similar conditions prevail in Syria. Dr. A. Bonné, of the Jewish Agency Economic Research Institute in Jerusalem, states, on the basis of estimates given him by the French mandatory government, that of the 78,000 square miles of Syria, about 23,000 are cultivable, but only about 6000 are really cultivated. Of the 2300 square miles of irrigable land, only one third, is irrigated. He concludes that the present population of three millions could easily be doubled.

With regard to Trans-Jordania we have the cautious view of the Palestine Royal Commission: "The area of Trans-Jordan is about 34,000 square miles and its present population is estimated at about 320,000 persons. Thus, while the country is almost three and a half times as big as Palestine, it contains about a quarter of its population.

"That population, moreover, includes a large number of Bedouin, whose pastoral life requires more land than would be needed if in the course of time they were to settle down to agriculture . . . . We are not in a position to fix even an approximate figure for the possibilities of new settlement in Trans-Jordan; but in view of the evidence given by some of those who are acquainted with the country and from what we saw, we consider the hope justified that, if fully developed, it could hold a much larger population than it does at present."

What difference will the acquisition of Palestine or the ruin of Palestinian economy make to those "desperately" underpopulated Arab countries? What possible benefit can the poverty-stricken fellahin of Egypt and the Bedouin of Iraq derive from a stoppage of Jewish immigration into Palestine, or reversely, what harm can come to them if in that distant and foreign land of Palestine the Jews were to be granted political autonomy?

The *sine qua non* of Iraq's progress, says the British Government's official report, is an increase in population; and "any substantial increase of population in the near future must come from the outside."

If that then is the case, the matter is clear: Iraq is "gravely" underpopulated, but Egypt is "dangerously" overpopulated.

The total area of Egypt, including the Libyan, or Eastern, desert and the Sinai Peninsula, is 383,000 square miles. The cultivated area in the Nile Valley, the Delta, and various oases is 13,600 square miles. According to the census of 1937, Egypt has a population of 15,920,703 or nearly 16 millions. The administrative division known as "Governorates," with an area of 318 square miles, has a population of 2,583,613, *or 7,417 persons per square mile*. In lower Egypt, which has an area of 8,483 square miles, live 7,138,678 souls, that is, 842 people per square mile. Upper Egypt, comprising 4,773 square miles, has a population of 6,423,412, or 1,346 per square mile. *Thus the average density of population over the entire Egyptian kingdom is 1,173 persons per square mile*—a figure far in excess of that found in the most densely populated areas of Europe, such as Belgium and England.*

*These figures are from Statesman's Year Book, 1942.

Egypt has nearly 1,200 Arabs per square mile, as against seven Arabs per square mile in Iraq. Thus the density of the Egyptian population is nearly 170 times that of Iraq. And all of these Arabs are desperately poor.

Of the 12,000,000 adults in Egypt, according to the same British *Statesmen's Year-Book,* 10,500,000 are illiterate. Of the 1,500,000 literates 50,000 are Jews, 200,000 Greeks, and 250,000 Italians. This leaves about a million Egyptian Arabs who can read and write.

The agricultural population of Egypt, says the *Yearbook,* makes up 67 per cent of the population. Of these cultivators one half per cent own 37.01 per cent of all the cultivated land. They are the big landowners, or effendis. Ninety-nine and one half per cent of the cultivators are fellahin—peasants—who own together 34.4 per cent of the cultivable land. The other 28.59 per cent of the cultivable territory consists of crown domains. There is virtually no industry in Egypt.

Sixty-two per cent of Egypt's fellahin till farm plots of less than five acres. Twenty-eight per cent are landless. The relation between employers and employees on the big estates is hereditary. An effendi inherits not only the land from his ancestors but also the people employed on the estate. This is the feudal system pure and simple: masters and chattel slaves, as it existed in Europe in the Dark Ages.

Now in what way would the stoppage of Jewish immigration into Palestine benefit the Egyptian people, that mass of serfs? The Egyptian king and his cabinet are reported to be solid with the other foreign Arab princes in demanding that Great Britain put a stop to building the Jewish national home. Is that to benefit the Arabs of Palestine? But the Arabs of Palestine are a hundred times better off than the Arabs of Egypt. What is the game?

The game is to set the Jewish people and the Egyptian people at loggerheads; the game is: *divide et impera.* It is not enough that Palestine has been made to rankle in the heart of every Iraqi Arab. It must also be made to rankle in the heart of every Egyptian Arab. In such a way there will be a nice majority of ranklers at those Pan-Arab Conferences, which the British Agency and the Colonial Office convoke in Cairo. The Colonial Office itself will not need to say a word there about shutting down the Jewish national enterprise in Palestine. It will act, democratically at that, by virtue of and in obedience to the expressed will of the foreign Arab princes. The Jewish people will be outvoted.

The game is that when Farouk I or Nachas Pasha, his Prime Minister, speaks about Palestine (of which country the ten and a half million illiterates of Egypt do not even know the name), the voice is Jacob's voice, but the hand is Esau's! And Esau's address is not Cairo, Egypt, but Whitehall, London, England, the home of the Colonial Office. The Jews, and especially the Zionists in America and Palestine, are falling into an artfully set trap when they polemize with King Farouk or with King Ibn Sa'ud on the subject of Palestine. The argument lies not with these gentlemen. They are mere automatons. The men who confuse the issues are the servants of British imperialism and their *Genossen* in Britain, Egypt, and America.

If Great Britain is really so deeply concerned about the poor Arabs—and God knows they are poor—let her look at Egypt, where she has now ruled for nearly half a century and still has those appalling conditions of ten thousand illiterate, half-starved Arab slaves on every inhabited square mile. Let her look, or rather let a commission of neutral, independent men look at Egypt, where Britain has proposed and disposed these many years, and then let the commission look at Palestine, where the Jews have by indirect action nearly wiped out Arab poverty. And then let the commission report to the world, to America, and to Their Majesties Ibn Sa'ud and Farouk. Maybe then some eyes will be opened!

Iraq can take in the ten million surplus population from Egypt without difficulty. The advance guard of these immigrants could be set to work on restoring the marvelous irrigation system of Iraq's two immense rivers, the Tigris and Euphrates, which collapsed after the Mongol invasions of the thirteenth and fourteenth centuries. For five hundred years no government in Iraq has been able to maintain the canals or regulate the flow of water. Many canals have silted up. Other canals, cut in the banks by the cultivators, without adequate knowledge or regard to the future, came to carry more water than the parent river. In many cases, the abstracted water, because of improper drainage, has ruined the land through salination or else filled vast marshes. The rivers, unable to scour their beds in the lower reaches, have increased the dangers from floods and hampered navigation.

Not until modern times were efforts made by the Turks, says Dr. Philip Ireland in his standard work on Iraq, to study methods of irrigation and to undertake irrigation works. The Turks, however, lacking money and initiative to carry out the plans of the British engineering corps that studied the situation under Sir William Willcocks, did undertake the construction of the Hindiya barrage and the Habbanya escape, but left the work uncompleted.

Subsequently, the British returned to the task. Since the occupation of Iraq commissions of engineers have roamed the land. Military engineers have taken the most urgent repairs in hand. Plans were submitted to complete the canals left half finished by the Turks, dig new ones, restore the pre-Mogul ones, and import and advance to the cultivators plow cattle and seeds. The schemes were but partially successful. They always broke down on what the British High Commissioner and other officials called "shortage of labor alike for construction and cultivation." There was no use, they said, opening up new areas for cultivation if there was no increase in population.

But the plans are there, worked out in minute detail. Cultivable land for all the roaming nomads of Iraq and the Nejd and for millions more can be provided. In recent years the experience of the Dutch in desalining or leeching the reclaimed land of the former Zuider Zee has increased tenfold the already vast possibilities of agriculture in Iraq.

Let us imagine for a moment that instead of giving the destitute and hungry masses of Iraq and Egypt stones for bread in the form of political pie in the sky, dreams of empire, and visions of loot in distant Palestine, the British Empire, in gratitude for having itself been plucked as a firebrand from the fire, were at the close of the war to proclaim the holy year of jubilee, turn up a new road, and let compassion and respect for the human dignity of "lesser breeds" be the guiding principles of its colonial policy! What new splendor would then descend on England! Then, indeed, her name of mother of the free would shine with new luster! Instead of playing the game of guile and blood, "to be called in righteousness, to be a covenant for the peoples, a light unto the Gentiles, to open the blind eyes, to bring out the prisoners and them that sit in darkness out of the prison house"!

What immense and exalted destiny!

If Britain would, upon the termination of hostilities, let the dead bury the dead and consider the Near East a blank sheet of paper, what great and beautiful characters could she trace on that page in collaboration with the two gifted and noble Semitic peoples!

Iraq's ancient and ingenious irrigation system repaired and her land redeemed!

Instead of empty spaces and desolation the broad valleys and sloping hills of Mesopotamia, the site of the legendary Garden of Eden, settled with a prosperous Arab peasantry ten or twenty million strong!

Fields of cotton where now the jackal roams and the lizard crawls!

Forests of beech, elm, cypress and eucalyptus in the now empty and dying soil and in the soggy marsh!

Wheat fields, olive groves, vineyards, orange plantations, honey hives, dairy farms, flour mills in place of mud hovels and human settlements built from castaway gasoline cans!

Food for India's starving millions grown right at that country's door and the abject, inhuman crowding in that country somewhat relieved!

The Egyptian land freed from its last and greatest plague: overpopulation!

Palestine's poorest Arabs freed from the slavery of their feudal lords and the bondage of Arab usurers and moneylenders!

The nomad, after wandering and being driven by hunger since times immemorial, sitting at last under his own fig tree, by his own hearth, in his own house, surrounded by his children!

Schools in the place of heaps of camel dung!

Clinics in the place of blindness, anemia, rachitis, hookworm, scurvy, and the frightful infant and maternal mortality of present-day Iraq and Egypt!

Prosperous villages where now wander Bedouin goatherds!

Prosperity and peace in the land of blood feuds and tribal wars!

In this task there lies before Britain an opportunity such as Providence has never conferred upon any nation in the history of mankind.

Then, too, Palestine would come into its own as the industrial center, the laboratory, and the assembly plant for the new world awakening in East Africa and the immense new markets which the Second World War has thrown into Britain's lap. Britain's sway now extends unchallenged and undisturbed from the Cape to Cairo. Virtually the whole of the African continent is in her hands. The fantastic dreams of Cecil Rhodes and other empire builders are becoming a reality. Many an empire in East Africa alone await the work of man: Egypt, the Sudan, Kenya, Uganda, Tanganyika, Eritrea, Somaliland, Ethiopia, Rhodesia, Bechuanaland, the Union of South Africa—untapped, unexploited, unexplored.

This new empire must be developed if Britain is to compete with the United States, which is about to enter the foreign market with an industrial apparatus of unparalleled magnitude and efficiency. Here Palestine enters upon the scene with new and surpassing significance.

The supremely important role of Palestine (the superlative is by no means exaggerated) in a new, regenerated capitalistic global exploitation of the world's markets and resources, in which Great Britain and the United States are to be the two chief collaborators and competitors, is sketched in broad outline in a confidential memorandum drawn up by a liberal British statesman and circulated privately among leaders of the inner cabinets of the Empire and the great self-governing Dominions of Canada, New Zealand, Australia, and South Africa. The document is a guide to the future imperial world policy for the Empire. A copy of it came into my hands, not quite by accident, in May, 1943.

While it is necessarily technical, the document does indicate, it seems to me, that under a progressive capitalism such as we may expect to emerge after the war, Palestine is to become one of the most important factors in the scientific exploitation of the riches of Africa and, indirectly, is to have a great influence on the modernization of the Arab world. The document surely does place Palestine in an entirely new light and in a new position, not politically or militarily this time, but economically. In fact, it points to Palestine as one of the keys to the solution of the world problems of tomorrow.

I quote some excerpts from this document:

> The development of industry, especially of the chemical industry, after this war will probably take place on lines quite different from those on which the chemical industry has developed hitherto. The sources of heavy metals required in chemical manufacturing being depleted with ever-increasing speed and the pace at which the available oil wells are being exploited have led to pessimistic predictions on the part of geologists. Schemes for the conservation of mineral oil have been developed in the United States, although they are not yet strictly enforced. But in spite of that, a view is current that the American oil supply will not outlast the next fifty years. At any rate, a shortage may be experienced before that.* There is, of course, a supply of oil from the Middle East, the Dutch East Indies, and South America, but as oil will be used in ever increasing degree, not only as fuel but also as a starting point for many synthetic chemicals, it is important to realize that in using oil for all those purposes the nations draw

****The New York Times* reported on August 22, 1943, that vast oil operations by American firms are about to open in Saoudia-Arabia on the Persian Gulf.**

on a capital definitely limited in quantity.

During the Second World War British industry has centered its attention on the production of a limited number of important war materials of very high quality such as airplanes, guns, and tanks, which, however, will not be required after the war in the same quantities as at present. The British government has not encouraged new enterprises even if they appeared to be of value for the further development of the industry. Quite to the contrary, the declared policy was to leave the chemical production, e.g., of aviation fuel, synthetic rubber, aromatic hydrocarbons, and plastics, to American industry, on the ground that the raw material for such production—namely oil—is abundant in America and that it would be easier to import the finished goods than the raw materials. Furthermore, in the first period of the lend-lease transactions, the British government has paid for supplies with holdings British subjects had in American enterprises. The participation British industry had in certain important sections of American industry has therefore entirely disappeared.

After this war British industry will be faced with a very large and very powerful American chemical and mechanical industry. Britain will not be able to compete with such a well-equipped industrial power. *She will, most certainly, lose the South American market after the war, which will come under American economic domination. The Chinese market may also well pass into American hands. At least, America, which must participate in the reconquest of the Asiatic world, is not going to withdraw from that area with empty hands.*

One of Britain's greatest assets has been the fact that practically all the natural rubber the world required was produced within the territories of the British Empire. There can be no doubt that natural rubber will after this war be replaced by the synthetic product, and even the investment will have lost a great deal of its value.

What trends of development in the future chemical industry are discernible now? First of all, the importance of synthetic products will go on increasing, and even some metals will gradually be replaced by synthetic organic products such as plastics. One must therefore look for a source of organic matter which can replace oil. Such a source is undoubtedly carbohydrates—sugar, starch, cellulose. It is practically inexhaustible because carbohydrates are reproduced in the vegetable world every year at least once.

By fermentation and relatively simple chemical methods of conversion carbohydrates can be transformed into practically all the substances the industry requires. In the case of alcohol, this has been known for a long time; it has been converted on a very large scale into a number of important industrial products. But other types of fermentation appear to offer even more interesting possibilities for the synthesis of various industrial materials than ordinary fermentation—for instance, acetone butyl alcohol fermentation. This process has been used on fairly large scale in many countries because of the industrial importance of acetone and butyl alcohol in themselves. During the last years, however, methods have been devised to convert these two substances into a large number of chemicals, and further study will certainly open new avenues.

In the following lines a brief survey is given of the synthetic possibilities of acetone and butyl alcohol from the studies of engineers and chemists, who are aware of the fantastic development the postwar world may see.

The fact that butyl alcohol is converted by a very simple process into butylene shows the potentialities of butyl alcohol. Butylene is the basis for high-octane fuel and for the synthetic rubber of the Buna type. Butylene so far has been produced from the cracking of gases formed from petroleum at high temperature; it occurs in these gases as a minor constituent, and its isolation is a cumbersome and costly procedure. Butyl alcohol gives directly chemically pure butylene, and all the processes which have been based on cracking butylene are as well or better carried out with fermentation butylene. This applies specifically to the conversion into butadiene and synthetic rubber. Apart from aviation fuel and Buna rubber, other important substances can be made from butyl alcohol. Butylene can be converted by catalytic processes into aromatic hydrocarbons, which lead to dyes and pharmaceutical products.

Acetone is used in large quantities in the manufacture of smokeless gunpowder. But there are manifold other uses of acetone. It has been shown—to mention only a few possibilities—that acetone can be converted into isoprene, which is the basic substance of natural rubber and which is destined to play an important part in the development of synthetic rubber superior to the natural kind. Acetone can be converted into a number of substances called ketones which are characterized, *inter alia,* by very high-octane numbers. These ketones, therefore, may be the best solution for the problem of finding a fuel for the most powerful airplanes of the future which will certainly tend to operate higher antiknock values than the present-day 110-octane fuel. Both from acetone and butyl alcohol, and also from other equally valuable fermentation products, a number of plastics can be made, and undoubtedly new ones can be developed, so that it will be possible to prepare synthetic materials from fermentation products, with any desired mechanical qualities of the type which have so far been confined to metallic materials. Fatty acids like acetic acid, butyric acid, and their anhydrides are made easily accessible from acetone.

One can indeed assume that in the not too distant future, carbohydrates will replace oil (and coal) as starting materials for the most vital requirements of our civilization.

If the British Empire pays attention to these possible developments, it can create for itself a position which will enable it to compete with

American industry. The material basis for such a scheme exists within the confines of the British Empire, where one finds an abundance of carbohydrates. The surplus production of wheat in Canada and Australia, of corn in South Africa, and of cane sugar in India and the West Indies has been for many years past a serious economic problem in these countries. This problem will disappear at once if these commodities are looked upon not as foodstuffs only, but are also considered as raw materials for a great industry.

The scope of the developments sketched above is so great indeed that it will be possible to utilize the vast quantities of carbohydrates which the African Continent* produces practically without human help and which are entirely wasted today, and it will give rise to a systematic development of agriculture in the African continent. As Africa will probably become the backbone of the British colonial empire after this war, such a development scheme is of political importance, and it is very doubtful whether there exists any other scheme of equal importance for the future of the Empire.

The greater part of the scheme mentioned in the memorandum is based on work already done in various laboratories in the United States and other parts of the world, but chiefly in . . . Palestine under the guidance of Dr. Chaim Weizmann, the renowned chemist who is the leader of the Jewish national cause. Weiz-

*The British Empire's lines, because of America's appearance in the Asiatic field, are going to be shortened. Britain will concentrate on the vast, untapped African Continent.

mann's farseeing genius has revealed him as one of the greatest empire builders of all times. The memorandum also states: "There is a definite prospect that the technical and economic feasibility of such a scheme will be tested first in Palestine, where the conditions for the realization of such a large-scale combined agricultural-industrial project are favorable."

Palestine can become, the memorandum shows, the laboratory and pilot plant for the big factory into which the African continent might eventually develop. There is no doubt that such a scheme will give Palestine a chance to settle a relatively large number of immigrants in a productive manner. At the same time, the link which will be created between Africa and Palestine may strengthen Palestine's position in the Arab world or even make it feasible for her to belong economically and politically to an African bloc, instead of entering the prospective Arab federation. The world of tomorrow will be organized on lines of economic regionalism rather than on political lines.

Dr. Walter C. Lowdermilk, assistant chief of the Soil Conservation Service of the United States Department of Agriculture, who is a great humanist as well as a scientist, will in a forthcoming book* give the result of his studies in the Holy Land and Arabia. He will show that boring a tunnel from the Mediterranean to the Dead Sea will give Palestine the highest waterfall in the world, unlimited electrical power, and a way of diverting the waters of the Jordan River to irrigation, thus increasing the productivity of the soil five times. Vast regions in the Holy Land moreover, such as the

*To be published by Harper's in 1943.

Negev and Trans-Jordan, are uninhabited, leaving room for hundreds of thousands of settlers and providing food, if properly cultivated, for a great industrial population, which must move in if the African continent is to take its place as it is intended to take its place, in a new economic order of world dimensions.

Certainly, in the light of the enormous industrial and economic flight that the British Empire will take in the next epoch, when the apex of capitalist civilization will be reached—for the earth has not yet been scratched, and the greatest era of capitalist prosperity still lies in the future—the political strife in the Holy Land sinks into the insignificance of the squabbles of washwomen around the village pump. Prosperity will make the Arab peoples forget the artificial political issues that have been injected into the situation to keep them occupied during the transition period.

Room and work and well-being there will be for all, for Arabs, Jews, and millions more. Political questions will sink into the background in the world of tomorrow when great economic unions and regional economic federations between nations come to the fore. If Britain and America enter in a competitive collaboration for the development of the world's markets, it will take imagination and energy and courage. If they shrink from the task and allow themselves to be diverted from their mission by fear of the Soviet Union and first combine in an effort to eliminate the Socialist state, they will plunge the world into chaos and into a series of bloody wars that will be the threshold of revolutions and the destruction of their economies . . . .

From before the dawn of recorded history, before this or that economic system grew up, mankind has carried in its bosom a predetermining sense of justice and right. It is this ideal that man has pursued from one age to another, from one form of civilization to the next. It is true that the quest was vague and groping, that man has not everywhere grown conscious of the road he travels and of the destiny to which he subconsciously aspires. There have been many pitfalls by the wayside and many obstacles in his path that have delayed the procession of the ages. But his ingenuity overcame all hindrances, often at the cost of blood and tears. The advance continued step by step—painful steps, stumbling, hesitating . . . .

Humanity moves. And when it moves it is not by and through the mechanical and automatic transformation of the modes of production, but under the now clearly, now obscurely felt influence of the beckoning ideal. The ideal itself, the idea is the principle of movement and of action. Instead of intellectual conceptions deriving from economic facts, it is the economic facts which translate and incorporate, in the reality of history, the ideal of humanity.

Slowly—it is a process of centuries, an immensely long accumulation of experience—humanity tears itself free from the period of the inconscient wherein it wandered during aeons of which we do not even know the length, for man has been on this earth more than a million years, and the last ten thousand years are but the portals of history. Human history is only beginning to dawn now. Humanity has been as a sleeping passenger on board of a ship drifting with the stream without contributing to the force of the current and awakening from time to time to see that the landscape has changed. He is awakening now. The landscape has changed abruptly.

Man has been pushed by the blind forces of events. But now a new era opens. Instead of being the inert object of elementary forces, man grows conscious of himself. He will henceforth himself regulate the march of events. He laughs at his childhood's fears of fatality and fate. The human community will be mistress of the great means of production, according to the known needs and requirements of men.

A new idea has been injected in the dormant consciousness of millions: the Four Freedoms, the right of every creature on earth to be free from want and free from fear. That idea is dynamite! Its explosion has awakened the world of Asia. Never again will the masses of men in Asia rest till they have made those freedoms their own. They may still be silently waiting for a time. But if their inalienable rights are not freely given to them, they will roar like a mighty cataract and sweep all before them. They will grow into a fire that is positive in that it transforms everything in its path into its own flaming substance.

At times the river of mankind courses serenely. In certain moments of history it breaks its dams and rolls forward with irresistible force. The dam has been blown up by the promulgation of the Atlantic Charter. That act can never be undone. The light of the ideal of justice grows vivid in our time!

For remedy of present evils and the prospect of conflict nothing adequate will be found till mankind is brought to that temper and liberality of mind which befit citizens of the world and till it feels indifferent, as one human race, as to whether industry is carried on in England, in India, in Palestine, or America—or in all places at the same time. The coming era will allow the great powers to behave handsomely and liberally and with essential kindness. They can afford to recognize the colonial peoples as equals. For only then will their collaboration be of value. India should be freed. The Jews must have a culturally and politically autonomous state of their own in Palestine. Only by being rooted in their own people and soil and character can individuals or nations develop their spiritual and intellectual inheritance to its highest significance and thus, as Gustav Streseman said once in the League Assembly, passing beyond its own race, be able to give something of themselves to humanity as a whole.

Winston Churchill has spoken but one word about Palestine in the course of this war. He has assured the Jewish people that he is confident that the ideals of the prophets of Israel will come through. That was excellent hope and noble encouragement. But it is not enough. We know that the ideals of the prophets of Israel *in abstracto* will never die and that the words spoken in Judea and Galilee, even if the Jewish people should perish and not leave a trace of their passage through history, will go ringing through the ages till the end of time . . . .

In his *Zarathustra,* Friedrich Nietzsche tells of the

return of the sage from his mountain of meditation where he had spent seven years. As he descended into the valley and entered the first village, he saw a crowd of people in the market place listening to a man who harangued them. Zarathustra approached and learned that the speaker was a ropedancer. The man had stretched a tightrope between two buildings and announced that he would presently give an exhibition of his art. But before carrying out the performance he talked of the difficulty of rope dancing. He told how nimble and agile one had to be to venture upon the trembling trapeze, how well balanced physically and mentally one had to be. He talked on and on and on and on until someone in the crowd impatiently and in exasperation called out: "We have now heard enough about tightrope walking. We would like to see some of it!"

We, too, we the common people of the world, the disinherited, the weary and heavy-laden, the Jews, the Indians, the Chinese, the Arabs, the Negroes, the colonial peoples, the prisoners of starvation, we, too, have heard enough about the ideals of the prophets of Israel, about the brotherhood of man, about equality of races and justice . . . . We would like to see some of it in practice. We would like to see some justice done!

The ideals of the prophets of Israel have been hovering in the unattainable clouds of abstraction long enough. Long enough they have been in heaven. They must be brought down to earth, out of the realm of metaphysical and philosophical speculation. The Word must become flesh. The ideals of the prophets must be translated into concrete human relationships, into new national and international covenants, into freedom for India, freedom for China, and freedom for Arabia, and into a Jewish State in Palestine!

Finished writing as the Angelus rang out from the tower of Christ Church in Bronxville, N. Y., Saturday evening, August 28, 1943.

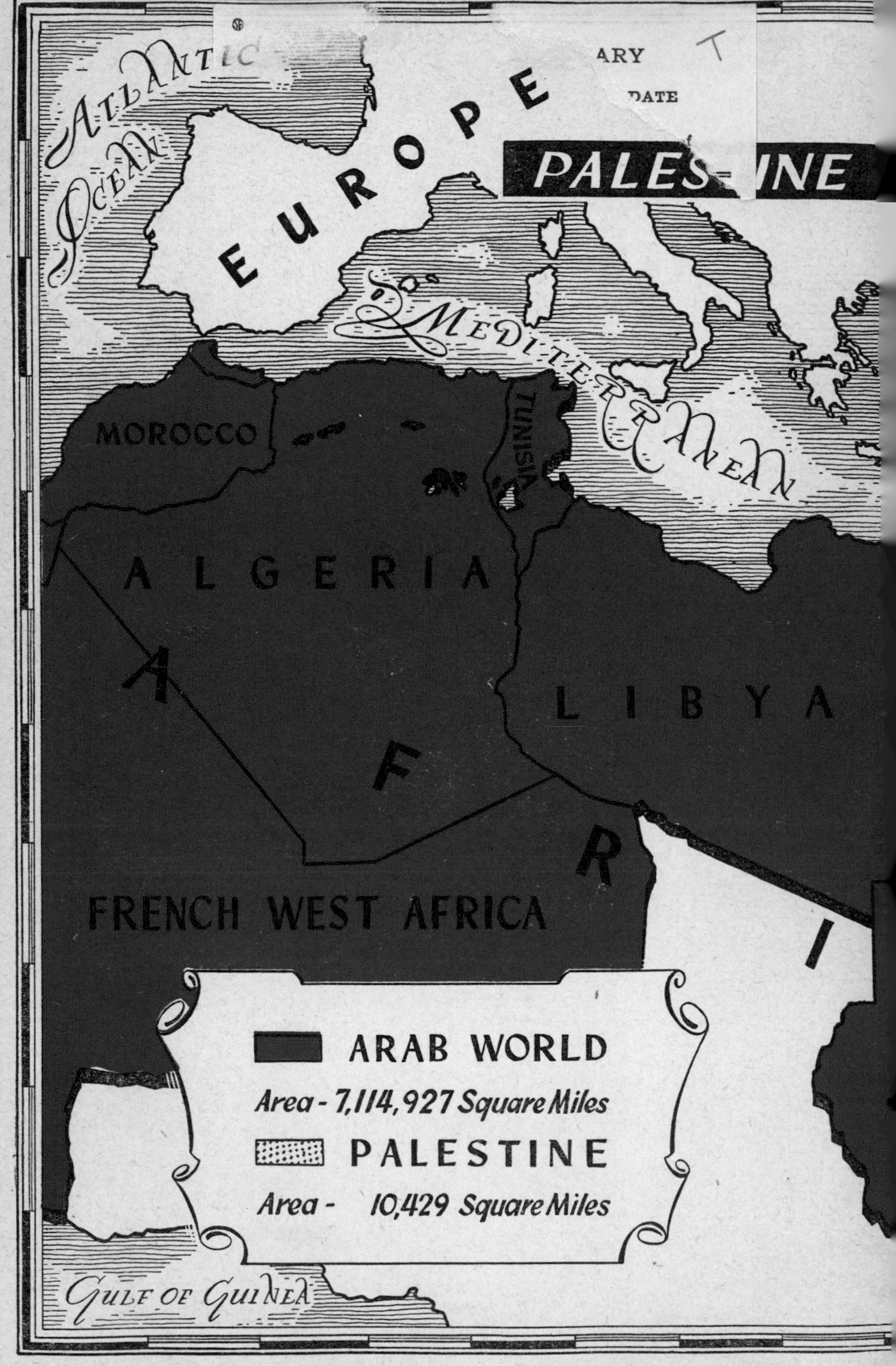
ATLANTIC OCEAN
EUROPE
PALES-INE
MEDITERRANEAN
MOROCCO
TUNISIA
ALGERIA
LIBYA
AFRI
FRENCH WEST AFRICA
ARAB WORLD
Area - 7,114,927 Square Miles
PALESTINE
Area - 10,429 Square Miles
GULF OF GUINEA